AMERICAN MILLING MACHINE BUILDERS 1820-1920

Kenneth L. Cope

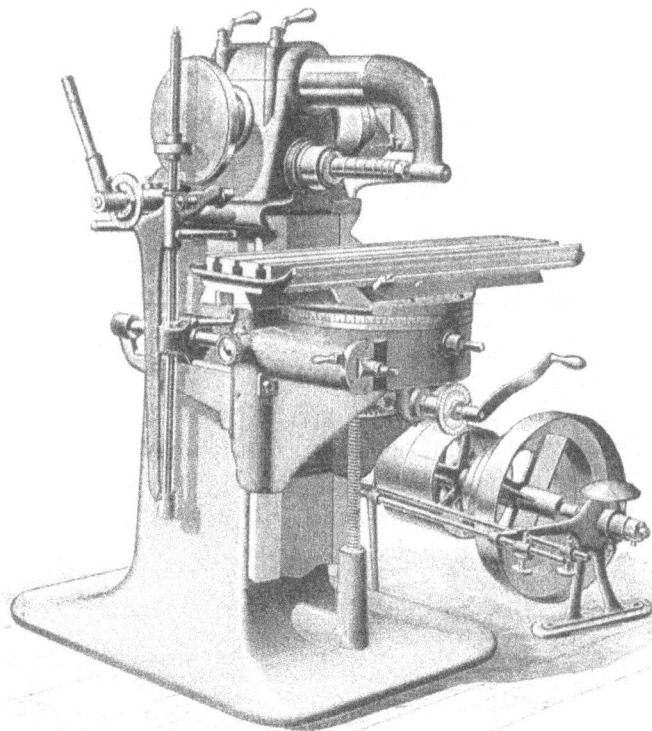

ASTRAGAL PRESS
Lakeville, Minnesota

Library of Congress Control Number 2007933788
International Standard Book Number 978-1-931626-24-8

Cover design by Donald Kahn

Published by
THE ASTRAGAL PRESS
An Imprint of Finney Company
8075 215th Street West
Lakeville, Minnesota 55044-8986
www.astragalpress.com

DEDICATION

Dedicated to the men and women, past and present, of the now almost defunct American machine tool industry. Often working under very difficult conditions, they made the tools that made the machines that made the United States the envy of the industrial world.

ACKNOWLEDGMENTS

Heartfelt thanks are extended to all those who helped make this book as complete as it is. Many people sent material from catalogs, city directories, and contemporary magazines from their collections or from sources to which they had access.

Special thanks to;

Vince Iorio, Annapolis, MD

Ken Kranzusch, Montgomery, AL

Frank Morrison, Fitchburg, MA

Phil Platt, Worcester, MA

Bob Vogel, Washington, DC

Also thanks to the staff of the Milwaukee Public Library and many other librarians scattered across the United States who were unfailingly polite and helpful when I called.

INTRODUCTION

Continuing the pattern set by the previous two machine tool studies, *AMERICAN LATHE BUILDERS 1810-1910* and *AMERICAN PLANER, SHAPER & SLOTTER BUILDERS 1830-1910*, this book is not meant to be a history of the milling machine. It is meant to bring to light the pioneer builders of American milling machines and to illustrate, as far as possible, their products. Many of the builders listed are well known, and only a very few are still operating; others are obscure, and some are known only from a single advertisement placed in a contemporary magazine.

Information and illustrations have been gleaned from such sources as city directories, records of sales to the U.S. government, comtemporary publications including *AMERICAN ARTISAN, AMERICAN MACHINIST, AMERICAN MANUFACTURER, ENGINEERING, IRON AGE, MACHINERY, MECHANICS* and *SCIENTIFIC AMERICAN* magazines, and a large number of catalogs issued by milling machine builders and dealers.

To put the entries into context, perhaps we need to start with short, late 19th century, definitions of milling machines and milling processes; the identification of the parts of a common type of milling machine; and some illustrations of milling machine operations:

DEFINITIONS

ANGLE MILLING—the machining of a surface at some other than a right angle to the axis of the milling cutter.

BENCH MILLING MACHINE—a small milling machine designed to be mounted on a bench and therefore of limited height and weight. Such machines are usually hand fed.

CONTINUOUS MILLING MACHINE—a milling machine designed to feed a continuous supply of workpieces into the cutters via a rotating plate, or dial, which can be loaded and unloaded while the machine is in operation.

DIE MILLING MACHINE—a form of vertical milling machine designed for cutting cavities in dies used in metal forming operations. Also called DIE-SINKING MACHINES.

DUPLEX MILLING MACHINE—a form of horizontal milling machine fitted with two, separately adjustable, milling spindle heads, usually mounted opposed at 180 degrees.

EDGING MACHINE—see PROFILIER

FACE MILLING—see SIDE MILLING

FORM MILLING—the machining of a special cross-section generally composed of straight lines and curves, or wholly of curves, with a milling cutter shaped to the desired form.

GROOVING—the cutting of grooves, or slots that may have any profile and follow a straight, helical, or irregular path.

HAND MILLING MACHINE—so named because the table or cutter is fed by hand. A typical design is arranged so that the table can be fed lengthwise by a hand lever. The spindle head or knee also has a vertical lever feed. Such machines are usually of small size and short travels.

KEYWAY MILLING MACHINE—a milling machine designed to cut keyways or slots, usually in shafts.

KNEE-TYPE MILLING MACHINE—a design with a vertical column which supports the cutter spindle and has on the face an adjustable knee on which the work table is mounted. Knee movement forms the vertical axis of motion.

LINCOLN MILLING MACHINE—named after George S. Lincoln, proprietor of the Phoenix Iron Works where the machine was developed about 1855. Machines of this type are intended for machining large numbers of duplicate parts and are not adapted to a great variety of milling operations. The work table, instead of being carried by an adjustable knee, is mounted on the solid bed of the machine and the outer arbor support is attached directly to the bed. Such machines came to be called MANUFACTURING MILLING MACHINES during the first two decades of the 20th century.

MANUFACTURING MILLING MACHINE—see LINCOLN MILLING MACHINE.

MILLING—the process of removing metal by a cutting tool that is rotated about its own axis and has one or more cutting edges that are successively brought against the work.

MULTISPINDLE MILLING MACHINE—machine fitted with two or more spindles that carry the cutting tools. Each spindle is usually made to be independently adjustable in relation to the work. In most such machines, the work can be moved in a straight line in one direction only.

PLAIN MILLING MACHINE—used for machining surfaces that require the motion of the work to be in a straight line during the cutting operation. They are arranged so that the work can be fed to the cutting tool, or vice versa, and also in two horizontal directions at right angles to each other. The axis of rotation of the cutter is usually horizontal.

PLANE MILLING—the machining of plain, flat, horizontal surfaces by means of cylindrical milling cutters whose length is usually much greater than their diameters. Also called SURFACE MILLING.

PLANER TYPE MILLING MACHINE—a machine with a cross rail on which is mounted one or more milling heads. Cuts are usually taken in one axis only as the table is moved under the spindle(s). Also called a SLABBING MILLER or PLANO-MILLING MACHINE.

PLANO-MILLING MACHINE—see PLANER TYPE MILLING MACHINE.

PROFILER—a form of milling machine designed for the machining of vertical edges of workpieces of irregular contour, generally done with an end mill mounted in a vertical spindle. The exact cutter path is usually determined by a templet or profile attached to the workpiece or the fixture supporting it. Also called EDGING MACHINE.

ROUTING—a milling operation in which the work is presented to the cutter and then guided by hand.

SIDE MILLING—the machining of vertical surfaces, or surfaces at right angles to the axis of the milling cutter. Also called FACE MILLING.

SLABBING MILLER—see PLANER TYPE MILLING MACHINE

SURFACE MILLING—see PLANE MILLING

THREAD MILLING MACHINE—a machine which appears to be a cross between a lathe and a milling machine. It was designed to cut threads with rotating milling cutters rather than the usual single-point tool used with lathes. Much of the development occurred during World War I when thread milling was found to be best for production of artillery shells.

UNIVERSAL MILLING MACHINE—so named because the numerous attachments furnished with them adapt these machines to a very wide range of work. The axis of rotation of the cutting tool is usually horizontal; the work can be fed to the cutting tool in a vertical direction, and also in two horizontal directions; the angle between the latter can be changed within the limits imposed by the design of the machine. By means of special devices the work can be rotated at the same time that it is moved longitudinally in a horizontal direction. This adapts the machine for the production of helical and spiral work. One of the fewmachine tool types developed in the United States, it was invented by Joseph R. Brown in 1861-1862.

VERTICAL MILLING MACHINE—a machine in which the axis of rotation of the cutting tool is vertical. In some cases the machine is arranged so that the work can be revolved in a horizontal plane for the purpose of finishing circular surfaces.

WORM MILLING MACHINE—a form of thread milling machine designed for cutting worm gears.

IDENTIFICATION OF MILLING MACHINE PARTS

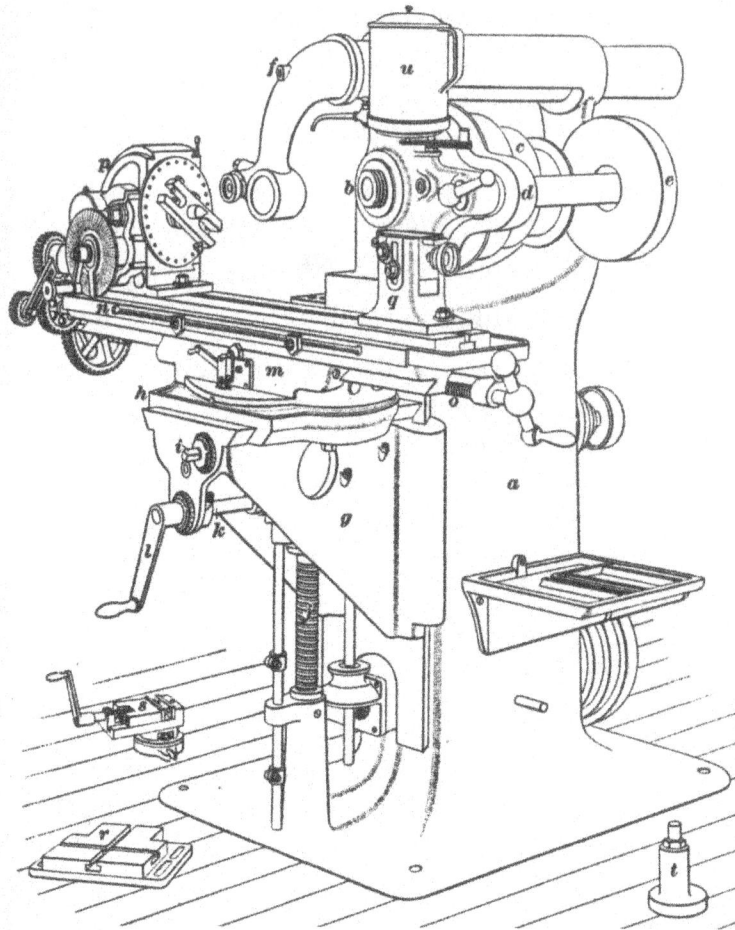

A—machine column,
    also called frame or standard

B—spindle

C—cone pulley

D—back gear

E—rear back gear

F—overarm

G—knee which supports the table and saddle

H—clamp bed or top of the knee to which the rotating saddle of a universal milling machine is clamped.

I—cross-feed screw

J—elevating screw

K—shaft to transmit crank motion to elevating screw

L—hand crank

M—saddle

N—table

O—table feed screw

P—index head, also called dividing head

Q—tailstock for index head

R—raising block

S—milling vise

T—steady rest to support workpiece on the table

U—oil tank for cutting fluid

MILLING MACHINE OPERATIONS

These illustrations show common milling machine operations as practiced at the turn of the 20th century. Little, if anything, would be changed if we were to do the same operations on any of today's manual machines.

Milling a Spiral.

Cutting Bevel Gear.

Milling Square-Head Screws.

Milling Irregular Form

Milling End of Compound Rest.

Milling T Slots.

Milling and Slitting.

Slitting a Tailstock.

Milling a Cam.

Milling a Dovetail.

Milling Convex and Flat Surface.

Milling Groove for Cap.

MILLING MACHINE OPERATIONS.

ADAMS CO., Dubuque, IA

Formed in 1892 when Eugene Adams (1860-1952) and his brother Herbert Adams (1863-1945) bought and reorganized the Roberts & Langworth Iron Works. Eugene served as president; Herbert as secretary and treasurer. In 1896 Adams introduced a milling attachment for planers, patented January 4, 1898, by F.O. Farwell. The attachment could be mounted as a horizontal spindle (Fig.1) or vertical spindle (Fig.2). Adams advised customers "do not plane surfaces that should be milled, or mill surfaces that should be planed." The attachment was made for 25 years or more. In 1907, the company introduced the FARWELL quick change miller (Fig.3). A hand feed machine, it was equipped with a quill for 2½" in/out motion and the 12" vertical motion was in the spindle head rather than a knee. It claimed the machine was more rigid than a conventional knee type milling machine, but a strong case could be made for the opposite.

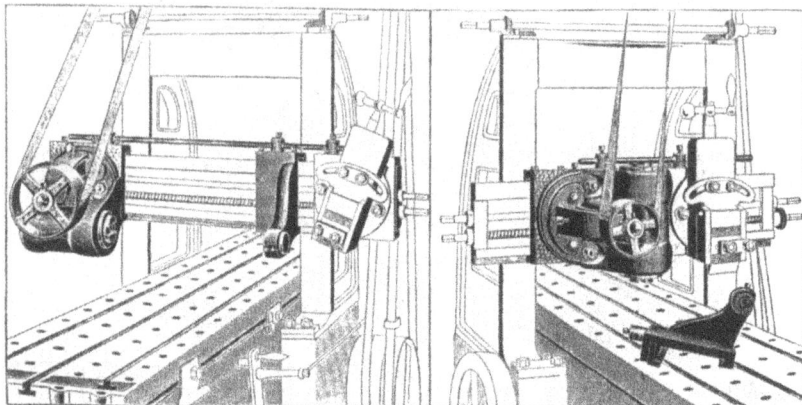

Fig. 1. Horizontal Spindle.   MILLING ATTACHMENT TO PLANER.   Fig. 2. Vertical Spindle.

Fig. 3

BUTT MILLING MACHINE BUILT BY JOHN ADT & SON, NEW HAVEN, CONN.

ADT & SON, JOHN, New Haven, CT

Builder of machinery for hardware manufacturers. Included was a butt milling machine, *shown left,* made for bench mounting. Note the dead weight used to power the feed.

AL-ED MFG. CO., Dayton, OH

Formed in 1910 to make bench milling machines. Very short lived, the company changed its name to the EXCELSIOR MACHINE TOOL CO. later in 1910.

AMERICAN GEAR CO., Yonkers, NY

Formed in 1886, the company made a variety of machine tools, including milling machines and lathes. No illustrations of its products have been located.

AMERICAN MACHINE TOOL CO., Hackettstown, NJ

Formed in 1915 to help meet the overwhelming demand for machine tools for European companies, and later for U.S. makers, who had contracts for European and American military material. 1915 production included Lincoln milling machines, *right,* made in No.1 size weighing 2,000 pounds and No.2 size weighing 2,775 pounds. Milling machine production probably ceased soon after 1918.

AMERICAN MACHINE WORKS, Springfield, MA

Maker of milling machines for Civil War gun makers.

AMERICAN MILLING MACHINE CO., Cincinnati, OH

Formed in 1913 to build a line of milling machines that, by 1920, included a No.1½ plain (Fig.1) and No.1½ universal (Fig.2) models along with a vertical milling attachment (Fig.3) that fit either style.

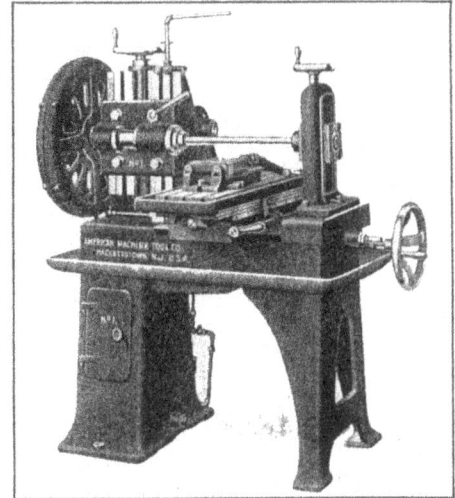

LINCOLN-TYPE MILLER

FIG. 1. AMERICAN NO. 1½ PLAIN BACK-GEARED MILLING MACHINE

FIG. 2. AMERICAN NO. 1½ UNIVERSAL BACK-GEARED MILLING MACHINE

FIG. 3. AMERICAN VERTICAL MILLING ATTACHMENT

AMERICAN MILLING MACHINE CO., Ludlow, KY

Absorbed in 1917 by the UNITED STATES MACHINE TOOL CO. which continued production of their products. See U.S. Machine Tool Co. entry for illustrations of the machines.

AMERICAN TOOL & MACHINE CO., Boston, MA

Formed in 1864 as the successor to the G.H. Fox & Co., the firm was best known for production of brassworkers' lathes. By 1893, the firm offered valve milling machines in two sizes, the larger of which, with 9" between centers, is *shown at right*. The company continued operations into the 1920s.

AMERICAN TOOL WORKS CO., Cincinnati, OH

Formed in 1898 by Franklin Alter (1831-1916), Albert B. Voorheis and Charles Davis (1853-1903) as a reorganization of the DAVIS & EGAN MACHINE TOOL CO. Capital was $1,000,000. The new firm continued production of Davis & Egan designs, including engine lathes, milling machines, drilling machines, shapers and planers. Milling machine production appears to have ended soon after 1900.

AMERICAN WATCH TOOL CO., Waltham, MA

A partnership of Ambrose Webster (1832-1894) and John E. Whitcomb formed in 1876 as a reorganization of Ballou & Whitcomb. The firm specialized in small machinery for the Waltham watch making industry and operated until 1918 when it was merged into the Wade-American Tool Co. Production in 1890 included the No.2 bench milling machine configured in several styles; as a dividing engine (Figs.1 & 2); a vertical swivel, vertical quill (Fig.3); vertical swivel, horizontal quill (Fig.4); bed slide with quill (Fig.5); and a horizontal swivel (Fig.6). A profiling machine (Fig.7) for pieces up to 2" square was also offered.

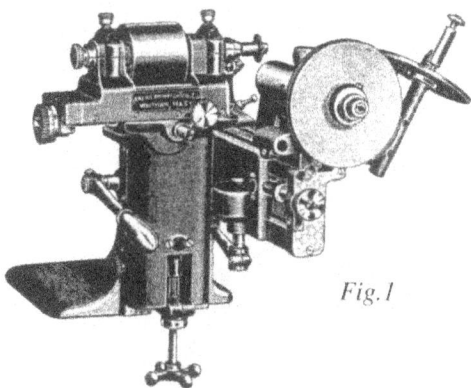

Fig.1

No. 2 MILLING MACHINE.
Dividing Engine—Left.

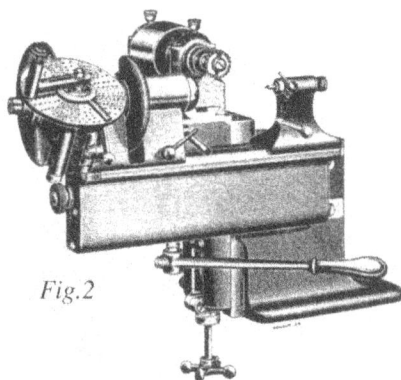

Fig.2

No. 2 MILLING MACHINE.
Dividing Engine—Right.

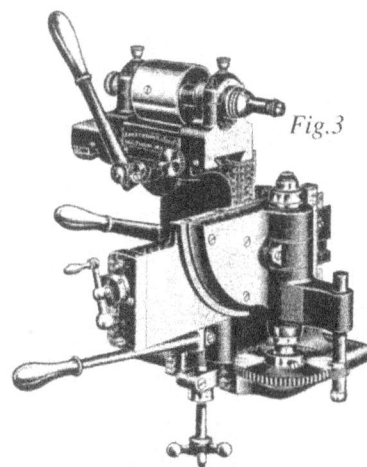

Fig.3

No. 2 MILLING MACHINE, 1-6 SIZE.
Vertical Swivel, Quill Vertical.

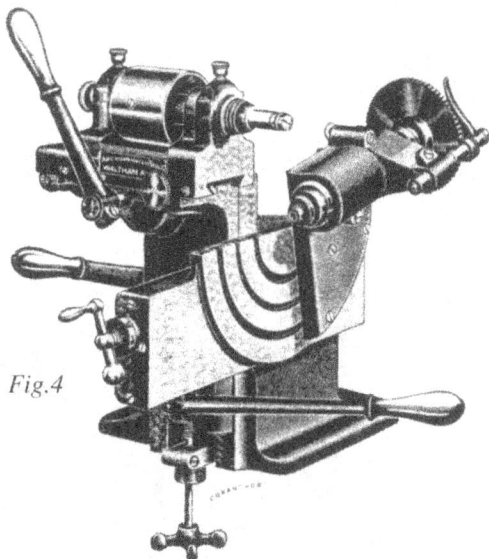

Fig.4

No. 2 MILLING MACHINE, 1-6 SIZE.
Vertical Swivel, Quill Horizontal.

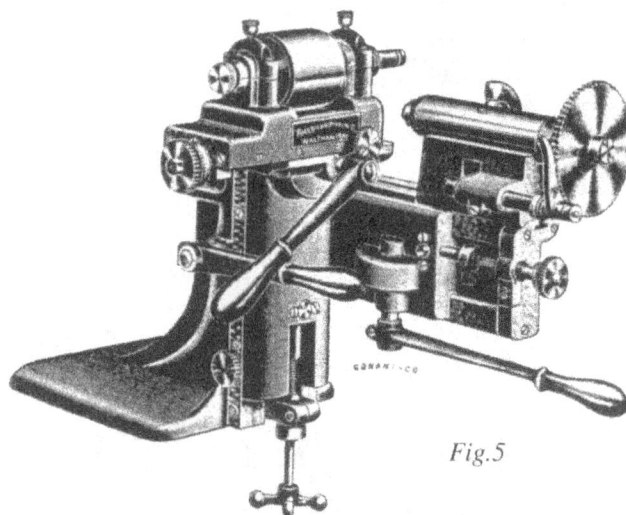

Fig.5

No. 2 MILLING MACHINE.
Bed Slide, Quill and Holder—Left.

(continued next page)

Fig.6

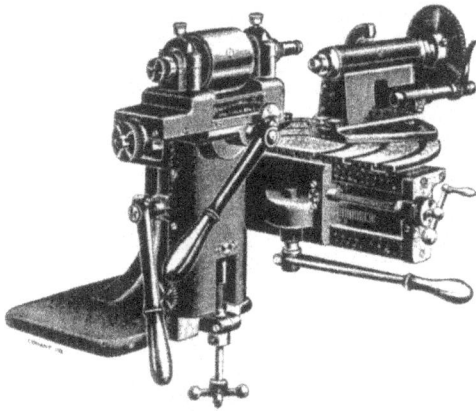

No. 2 MILLING MACHINE.
Horizontal Swivel—Left.

Fig.7

PROFILING MACHINE. 1-3 SIZE.

AMES CO., B.C., Waltham, MA

Formed by Bliss C. Ames (1867-1948) when the partnership of Alcorn & Ames was dissolved in 1901. Ames continued to make a line of small bench lathes and, by 1905, had added small milling machines. The bench millers, *shown at right,* were offered with dividing heads and made in one size only; 12" table travel, 3" cross travel and 7½" vertical travel. All machine production ceased in 1946.

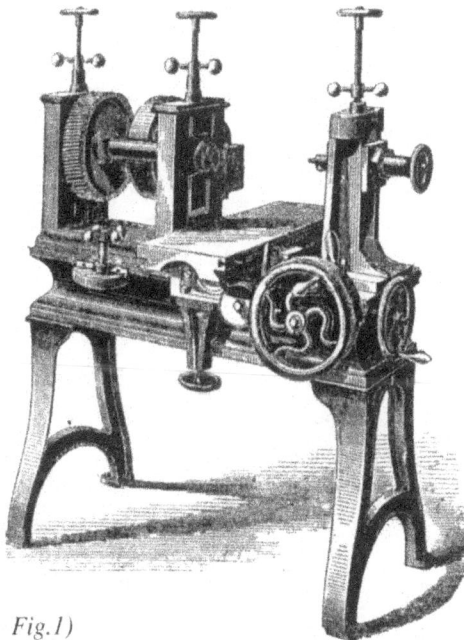

Fig.1)

THE M'FARLAND MILLING MACHINE.

AMES MFG. CO., Chicopee Falls, MA

Founded in 1834 by James T. Ames (1810-1883) and his brother Nathan P. Ames, Jr. (1803-1847) to make cotton machinery and machine tools. Ames was one of the first machine tool builders to make and market a standard line of machine tools. Milling machine production began about 1850 with the introduction of the MacFarland milling machine (Fig.1) designed by Jacob Cory MacFarland who worked at Ames from 1845 to 1858. The machine appears to be an improvement on the Howe miller made by ROBBINS & LAWRENCE. Vertical milling machines (Fig.2) were offered c1860-1870 and were furnished

Fig.2

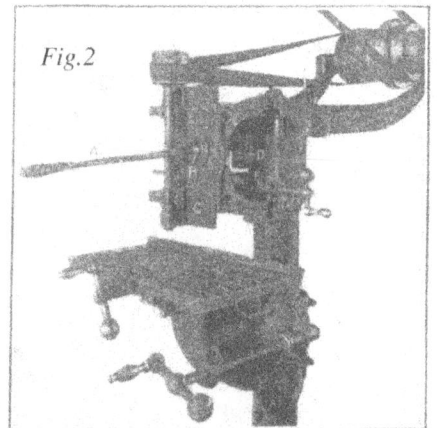

AN OLD DIE-SINKING MACHINE

11

with a lever spindle feed for use as a die sinker. From the 1850s through 1865 Ames was an important supplier of machine tools to the Springfield Armory. At the end of the Civil War, however, firearms, swords, sewing machines, etc. became important products. Machine tool production began to taper off and ceased soon after 1890. The firm was closed in 1898.

ATLAS MACHINE CO., Providence, RI

Maker in 1904 of milling machine vises made in tool maker's style (Fig.1) and as a plain swivel vise (Fig.2).

Fig.1

Fig. 1.—Tool Maker's Vise.

Fig.2

Fig. 2.—Plain Swivel Vise.

AURORA TOOL WORKS, Aurora, IN

Founded in 1889 by Alex B. Pattison. The firm specialized in upright and radial drilling machines but entered the milling machine market in 1902 with a line of plain millers. Milling machine production appears to have been short lived, ceasing by 1907. In 1925, the firm was bought by the Oesterlein Machine Co. which continued to operate it as a division for a short time.

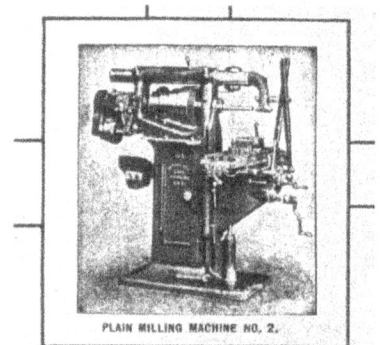

PLAIN MILLING MACHINE NO. 2.

AURORA MILLING MACHINES.

We build good plain millers; strong, rapid, convenient machines, designed to do good work at a minimum cost. Special circulars with description and general dimensions sent on request.

We also build Upright Drills, 14 to 40-inch.

Send for Catalog No. 7.

AURORA TOOL WORKS, Aurora, Ind.

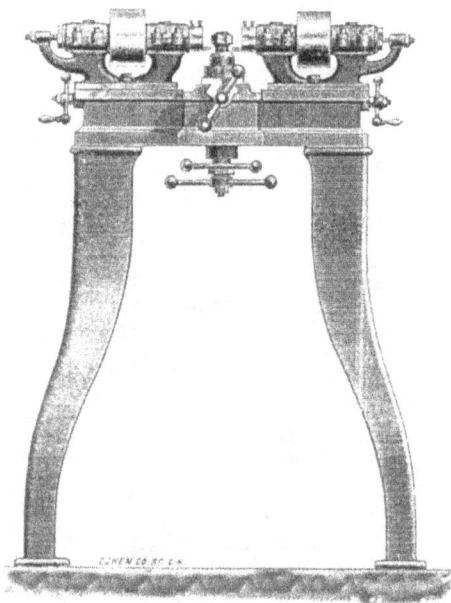

VALVE MILLING MACHINE.

BARKER & CO., WILLIAM, Cincinnati, OH

Formed in 1886 by William Barker (1853-1915) after selling his share of LODGE, BARKER & CO. Barker served as president until his death in 1915. His three sons then continued the business, probably through World War I. The firm made a variety of small machines for the valve industry including valve milling machines *(see left)* and cutting off machines offered in 1886. Machine tool production appears to have ceased about 1900. Barker was also a partner in BARKER & CHARD MACHINE TOOL CO. from 1897 to 1901.

BEAMAN & SMITH CO., Providence, RI

Formed in 1886 as a partnership of Elmer A. Beaman (1846-1921) and George H. Smith, both of whom had been working at the BROWN & SHARPE MFG. CO. Milling machines were the primary product but the firm also made a variety of lathes c1890-1891. Beaman served as treasurer and general manager until his death in 1921. Production centered around heavy duty milling machines that, in 1890, included a vertical spindle machine with 8' x 2' table (Fig.1), weighing 10,000 pounds; and a horizontal spindle machine (Fig.2) with a table 14" wide and 6' or 8' long; heavy Standard knee and ram type mills (Fig.3) with 72' x 14" table, weighing 8,000 pounds; and Standard milling machines (Fig.4) with an odd 25" x 32" vertical table," with the same ram and knee motion as the heavy Standard, but weighing only 5,500 pounds. *(continued next 4 pages)*

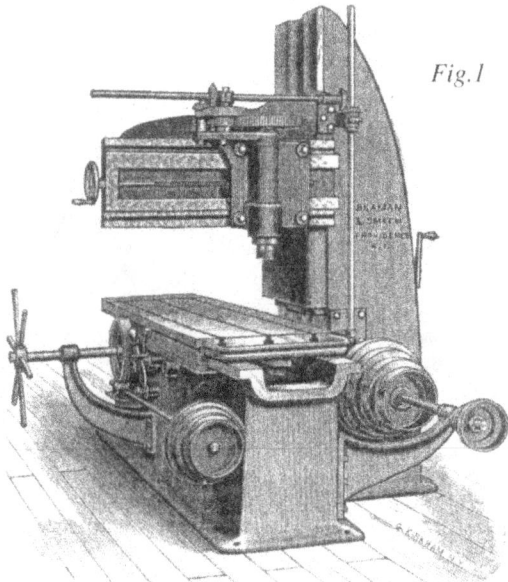

Fig.1

VERTICAL SPINDLE MILLING MACHINE.

Fig.2

HORIZONTAL SPINDLE MILLING MACHINE.

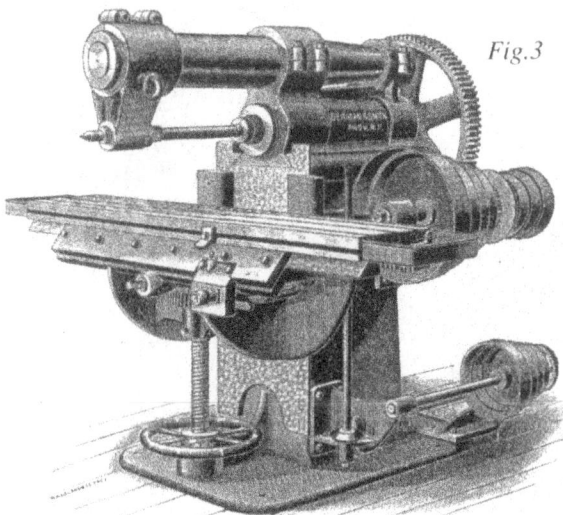

Fig.3

HEAVY STANDARD MILLING MACHINE.

Fig.4

1890 production also included a portable locomotive port milling machine (Fig.5) with 20" x 20" travel, later made by PEDRICK & AYER. By 1892, the firm offered Duplex milling machines in No.1 (Fig.6) size with a 40"x 27" table and no vertical travel; and No.2 (Fig.7) size with 54"x 14" table and 48" vertical travel in the knee.

Fig.5

LOCOMOTIVE CYLINDER PORT MILLING MACHINE.

Fig.6

DUPLEX MILLING MACHINE.

Fig.7

Fig.8

FOUR-SPINDLE MILLING MACHINE.

Fig.9

TWO-SPINDLE MILLING MACHINE.

Fig.10

A MANUFACTURER'S MILLING MACHINE.

1895 production included four-spindle milling machines (Fig.8) with two vertical and two horizontal spindles; and two-spindle milling machines (Fig.9) weighing 8,000 pounds.

Fig.11

DUPLEX MILLING MACHINE.

Fig.12

Horizontal Spindle Milling Machine, with Vertical Spindle.

Fig.13

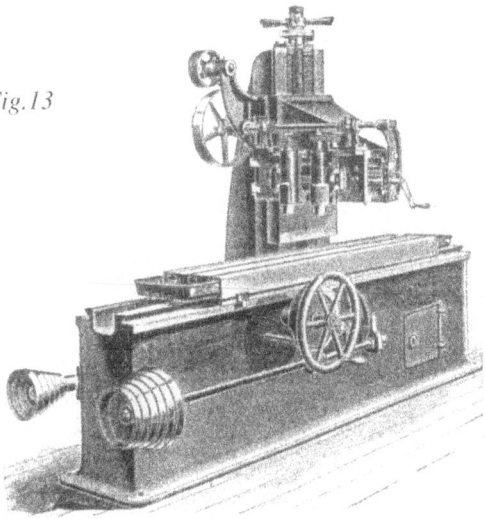

Fig.14

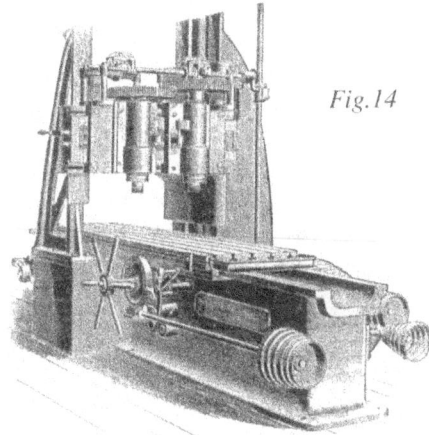

Fig.15

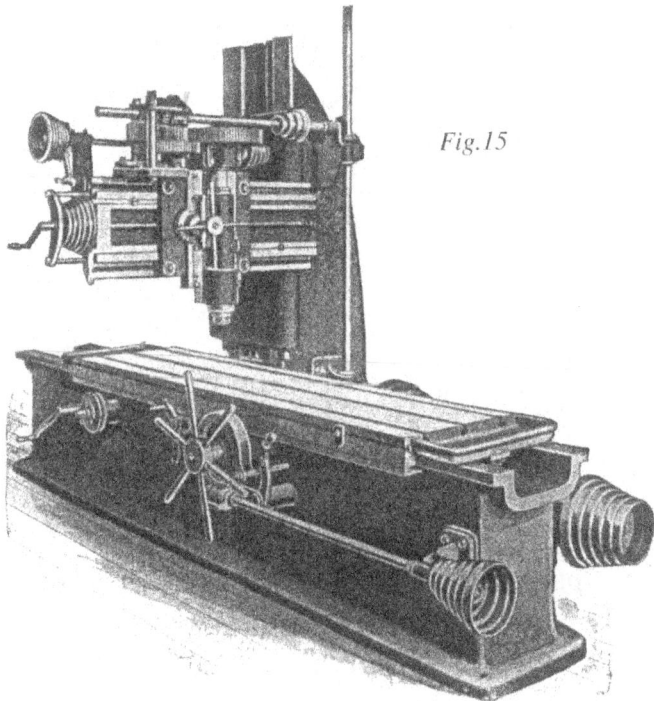

Fig.16

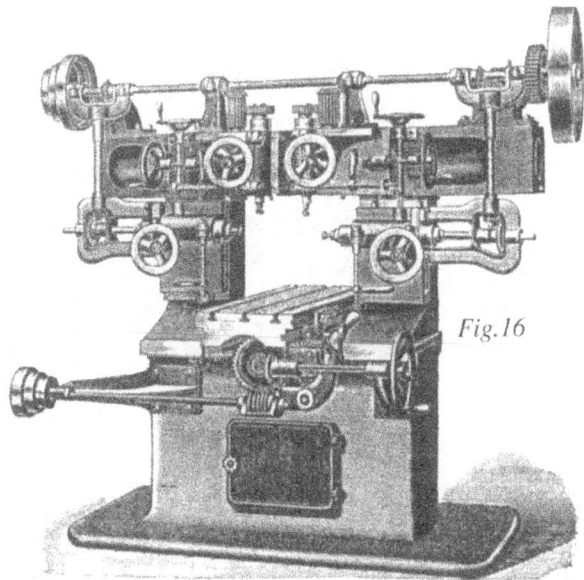

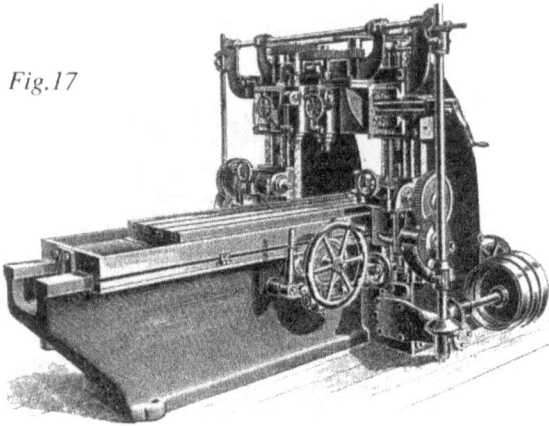

Fig.17

Fig.18

Fig.19

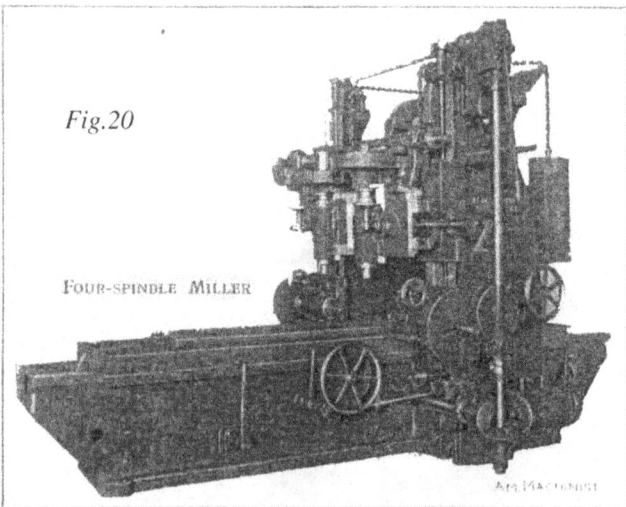

Fig.20

FOUR-SPINDLE MILLER

Heavy manufacturer's milling machines (Fig.10) were introduced in 1897, followed by duplex mills (Fig.11) weighing 20,000 pounds in 1898, and combination horizontal and vertical spindle mills (Fig.12) in 1899.

1901 production included No.1 vertical, two-spindle milling machines (Fig.13) weighing 18,000 pounds; No.2 vertical, two-spindle milling machines (Fig.14) weighing 20,000 pounds; No.4 vertical spindle milling machines (Fig.15) weighing 18,000 pounds; No.1 four-spindle milling machines (Fig.16) weighing 5,300 pounds; No.2 four-spindle milling machines (Fig.17) weighing 15,000 pounds; and No.2 horizontal spindle milling machines (Fig.18) weighing 14,000 pounds. A true duplex milling machine (Fig.19) with two separate heads and tables, was offered in 1903; and an updated four- spindle miller (Fig.20) in 1912. Boring and drilling machines, which had been made since about 1890, became the larger part of their business by World War I; milling machine production appears to have lessened from that point. Operation continued until 1927 when the company was absorbed by Manning, Maxwell & Moore.

BECKER-BRAINARD MILLING MACHINE CO., Hyde Park, MA

Formed in 1899 by the merger of the JOHN BECKER MFG. CO. and the BRAINARD MILLING MACHINE CO. Amos Brainard was president until retiring in 1901. John Becker, who had left the firm in 1903, returned in 1908 and reorganized as the BECKER MILLING MACHINE CO. The new company continued production of most of the milling machines previously made by Becker and Brainard. These included the Becker vertical machines in No.2, No.3, No.4, and No.5 sizes. A new No.5B size (Fig.1) was introduced in 1900. By 1903, ten sizes (Fig.2) were offered. Brainard designs that continued in production included the tool room universal model (Fig.3).

A new model of the Brainard No.7 Lincoln miller (Fig.4) with 20" table travel and weighing 1,770 pounds was introduced in 1902. 1903 production included heavy, planer type milling machines (Fig.5) with 120" travel, weighing 25,000 pounds; and an improved No.2 Lincoln milling machine (Fig.6) with 18" travel and weighing 1,750 pounds. A new line of knee type mills, based on Brainard designs, was introduced in 1901. These included the No.53 plain special (Fig.7), No.5B with 16" x 48" table (Fig.8), No.5C with 16" x 60" table, plain milling machines with angular/vertical milling attachment (Fig.9), and No.3 universal (Fig.10) milling machines.

(continued next two pages)

Fig.2

SIZES AND DIMENSIONS

No. of Machine	2	2A	3	3H	4	4H	5	5B	6	7
Working surface of platen......	17"x9"	18½"x9"	28"x10¼"	28"x10¼"	26"x10¼"	26"x10¼"	48"x13"	{43"x14" 53"x14"	{41"x18" 49"x18" 61"x18"	58"x26"
Longitudinal feed	12"	19"	21"	27"	28"	28"	39"	42" 54"	42" 50" 62"	60"
Cross feed......	12"	12"	12"	10"	13"	13"	14"	16"	20"	30"
Vertical feed of spindle	2"	3"	3½"	3½"	5¼"	5¼"	7"	9"	13"	15"
Max. dis. spindle to platen......	17"	17¼"	17½"	18"	19"	19"	20"	21½"	24"	26½"
Center of spindle to neck......	15"	15"	16"	16"	15"	15"	17"	18½"	24"	31"
Outside diameter rotary table..	15"	15¼"	18"	...	20"	20"	22"	22"	27¾"	36"
Speed of C. S. { fast........	700	400	400	400	400	400	400	400	450	320
{ slow........	125	125	100	100	100	100	100	100	275 100	200 75
Diameter of pulleys on C. S....	8"&12"	8"&12"	8"&12"	8"&12"	10"&14"	10"&14"	12"&18"	12"&18"	14",14",18"	18",18",22"
Width of belt..	2½"	2½"	3"	3"	3"	3"	3½"	3½"	4¼"	5½"
Net weight. ...	1,300	1,400	1,850	1,680	2,425	2,450	4,100	4,500	7,360	....

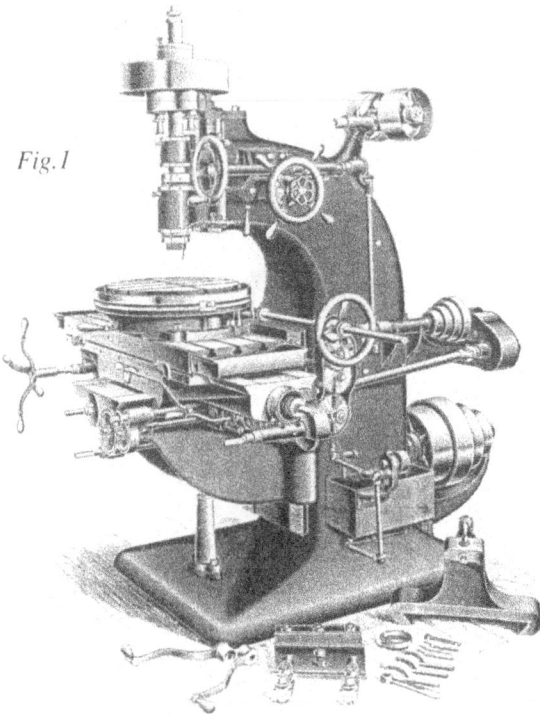

Fig.1

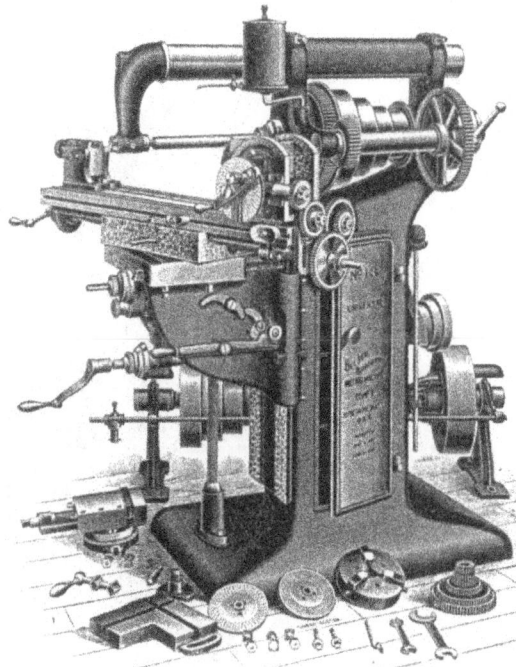

Fig.3

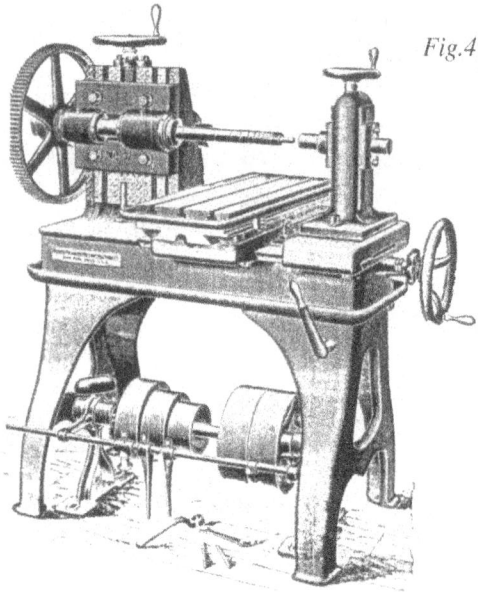

Fig.4

Fig.5

THE BECKER BRAINARD PLANER TYPE MILLING MACHINE

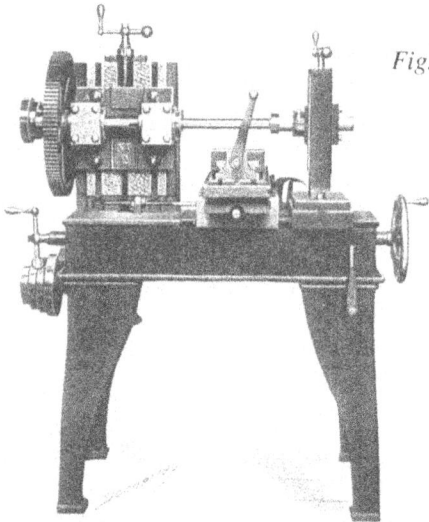

Fig.6

Fig.7

THE BECKER BRAINARD PLAIN MILLING MACHINE.

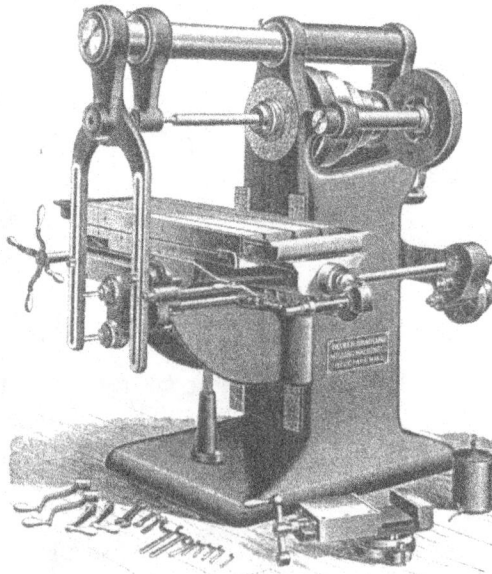

Fig8

Fig.9

THE BECKER-BRAINARD PLAIN MILLING MACHINE.

Fig.10

BECKER MFG. CO., JOHN, Boston, MA, later Fitchburg, MA

Formed in 1889 by John Becker (?-1927) to make vertical milling machines patented May 17, 1892, and January 3, 1893. Becker moved to Fitchburg, MA, in 1891. In 1899 he merged with the BRAINARD MILLING MACHINE CO. to form the BECKER-BRAINARD MILLING MACHINE CO. Becker's first milling machine (Fig.1), introduced in 1890, had a 8" x 16" table, spindle speeds to 10,000 RPM (a very high speed at that time), and weighed 1,100 pounds. It was later designated the No.2. The No.3 model (Fig.2), introduced in 1892, offered 16" table travel, 11" cross travel, 15½" vertical travel and weighed 1,500 pounds. It was available with an optional 9" chuck and later with an 18" rotary table. The No.4 model (Fig.3), introduced in 1893, was equipped with 28" table travel, 13" cross travel and a 20" diameter rotary table. *(continued next page)*

Fig.1

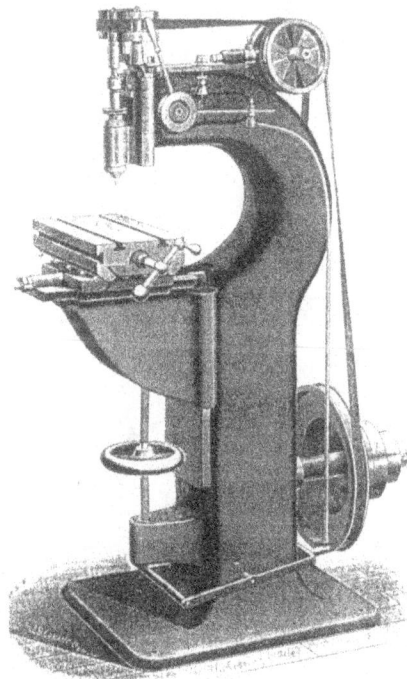

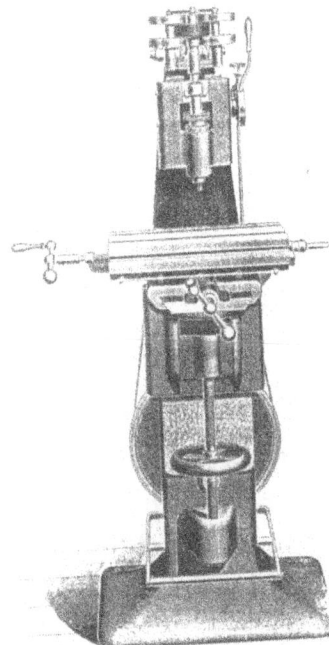

THE BECKER VERTICAL MILLING MACHINE.

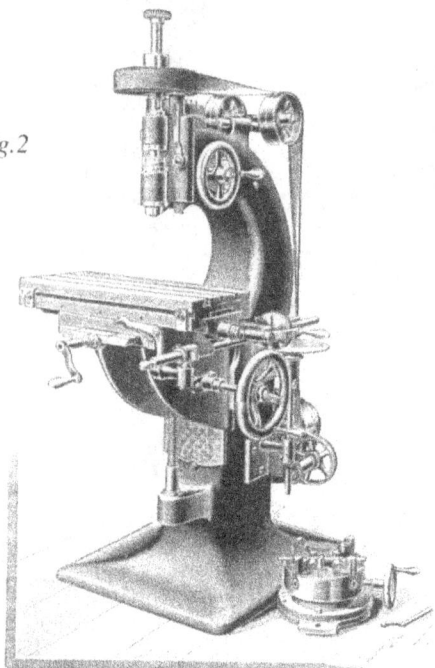

Becker Vertical Miller, No. 3.

Fig.2

Fig.3

IMPROVED VERTICAL MACHINE.

BECKER MILLING MACHINE CO., Hyde Park, MA

Formed in 1908 by John Becker (?-1927) shortly after returning to the BECKER-BRAINARD MILLING MACHINE CO. from which he had retired in 1903. He turned over active management to his son, John Becker, Jr., in 1914 and sold out in 1919. In 1923, the company was absorbed by the REED-PRENTICE CORP. The new firm continued production of the Becker vertical milling machines (Fig.1) in six sizes. A new AB model (Fig.2),

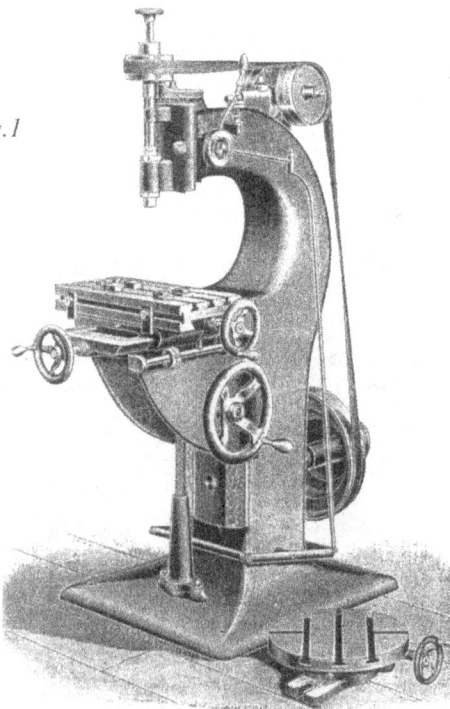

Fig.1

Fig.2

with a variable friction feed transmission was introduced in 1911 and a Becker model B (Fig.3) with 36" table travel and 14" cross travel in 1913. An improved No.7 Lincoln miller (Fig.4) was also introduced in 1913 along with the No.6 vertical (Fig.5) offered with 50" or 62" table travel. A high-powered, single-pulley-drive model (Fig.6) was introduced in 1914 and offered in four sizes of plain and four of universal millers. Oddly, the design included a complex belt drive to the spindle much like that used on the vertical machines. 1918 production included a new No.3 universal milling machine (Fig.7) with 30" table travel, 10" cross travel, and 19' vertical travel. Ads placed in 1920 (Fig.8) began to tout the great advantages of belt drive (most milling machines were gear drive by that time) and offered the new SD model vertical machine. *(continued next page)*

Fig.3

Fig.4

Fig.5

Fig.6

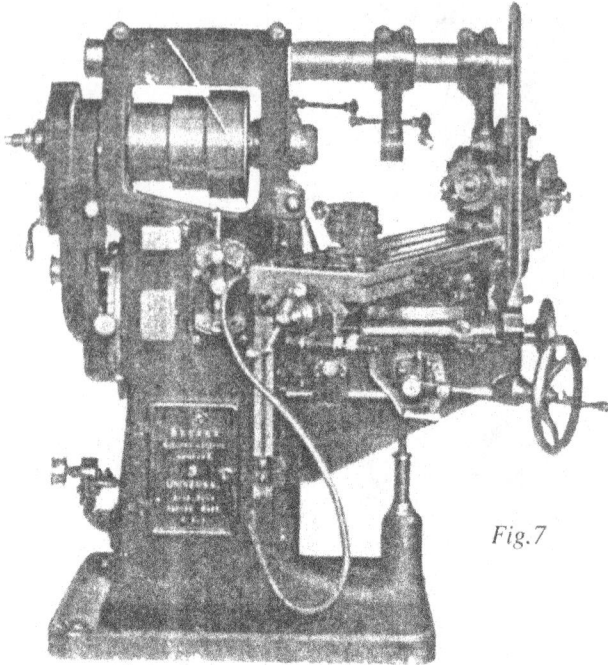

Fig.7

BEMENT & DOUGHERTY, Philadelphia, PA,

later **BEMENT & SON, WILLIAM B.**, Philadelphia, PA,

later **BEMENT, MILES & CO.**, Philadelphia, PA

A partnership of William B. Bement (1817-1897) and James Dougherty 1815-1900) formed in 1856 as a reorga-
nization of Bement, Dougherty & Thomas. Bement had been a noted designer of machine tools for the Lowell
Machine Shop 1845-1851. Products included vertical spindle milling machines (Fig.1), patented June 3, 1862, by
W.B. Bement. *(continued next three pages)*

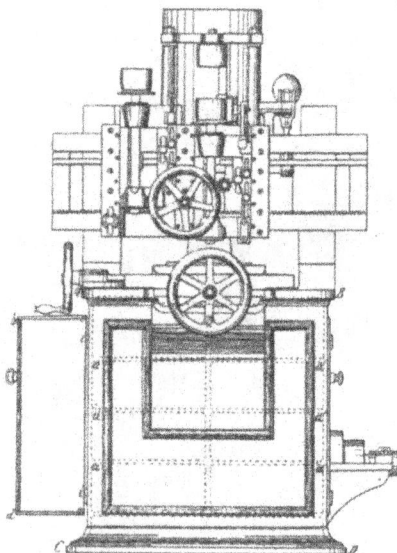

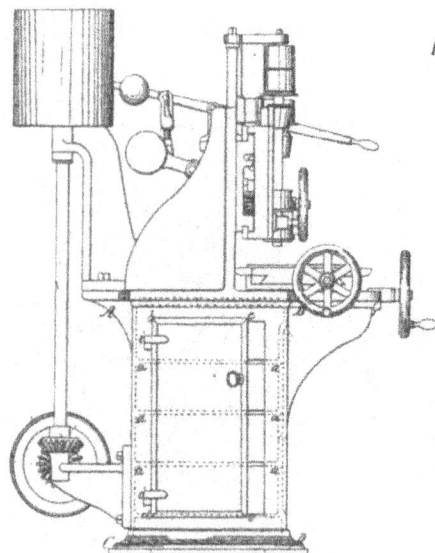

Fig.1

In 1870 Bement and his son Clarence S. Bement (1843-1923) formed a new company, William B. Bement & Son, which operated as the proprietor of the Industrial Works. Milling machine production included a model with an adjustable spindle head (Fig.2) and a type with a fixed spindle head (Fig.3). Both were introduced by 1876 and saw a wide sale.

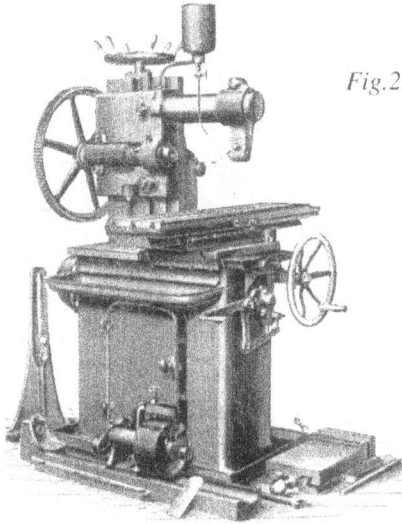

Fig.2

Fig.3

The firm reorganized as Bement, Miles & Co. May 1, 1885, when it merged with Frederick B. Miles' Machine Tool Works. Production of the earlier model milling machines continued for some time. A new vertical milling machine (Fig.4) with 42" rotary table was introduced in 1887 and a heavier model horizontal milling machine (Fig.5) was advertised by 1888.

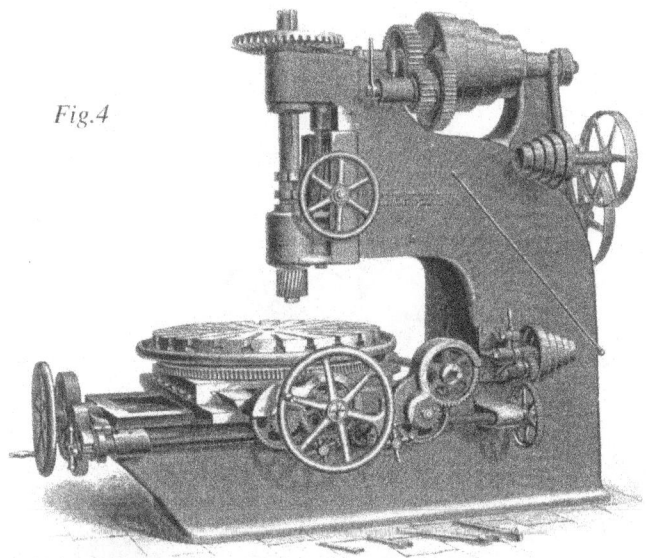

Fig.4

BEMENT, MILES & CO.,
PHILADELPHIA, PA.
——BUILDERS OF——
METAL-WORKING MACHINE TOOLS
FOR

Railroad Shops, Locomotive and Car Builders, Machine Shops,
Rolling Mills, Steam Forges, Ship Yards, Boiler Shops,
Bridge Works, Etc., Etc.

Fig.5

1897 production included newly introduced single spindle horizontal milling machines (Fig.6), with 24" between uprights, and a double spindle version (Fig.7) with 54" between uprights, both equipped with the Sellers friction drive.

Fig.6

SINGLE SPINDLE HORIZONTAL MILLING MACHINE.

Fig.7

DOUBLE HORIZONTAL MILLING MACHINE.

On August 15, 1899, the company merged with the NILES TOOL WORKS and POND MACHINE TOOL CO. to form the NILES-BEMENT-POND CO. Bement, Miles & Co. continued operation as a division. 1904 milling machine production included the No.5 double horizontal (Fig.8) with 42" between uprights; No.10 vertical with 42" rotary table (Fig.9); No.9 vertical with gear drive and 32" rotary table (Fig.10); No.9 vertical with belt drive and 32" rotary table (Fig.11); No.6 horizontal (Fig.12) with 37" between uprights; No.3 key-seating mill (Fig.13) which could cut slots up to 2½" wide and 16" deep; and No.4 vertical key-seating mill (Fig.14) which could cut slots up to ¾" wide and 2½" deep. *(continued next page)*

Fig.8

Fig.9

24

Fig.10

Fig.11

Fig.12

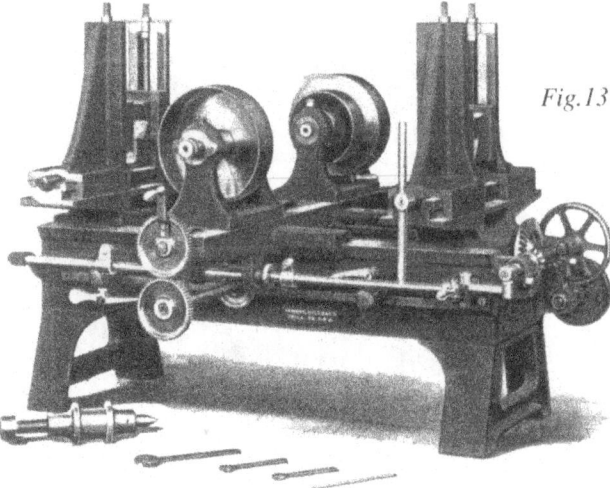

Fig.13

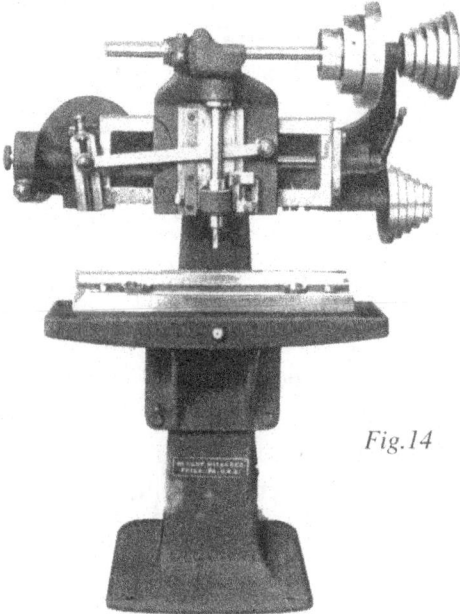

Fig.14

BEMIS, EDGAR W., Worcester, MA,

later **BEMIS MACHINE CO., E.W.**, Worcester, MA

Formed about 1909 to make a special milling machine designed to mill the hexagonal ends of valves, stuffing box nuts and nuts of all kinds up to 3" across the flats. The machine was of the dial index type where three pairs of cutters operated on each workpiece as it was indexed from a loading station through the three cutting stations. The firm was reorganized as the E.W. Bemis Machine Co. in 1919.

FIG. 1. BEMIS HEXAGONAL MILLING MACHINE

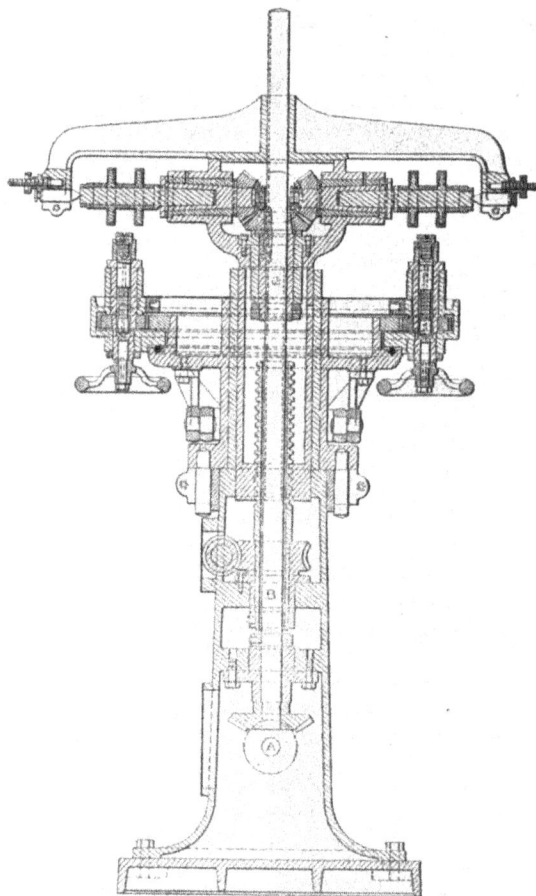

FIG. 2. CONSTRUCTION OF HEXAGONAL MILLING MACHINE

BETTS, E.& A., Wilmington, DE,

later **BETTS MACHINE CO.**, Wilmington, DE

A partnership of Edward T. Betts (1856-1916) and his brother Alfred Betts, formed in 1861 to make machine tools. The firm continued until 1878 when it was reorganized as the BETTS MACHINE CO., which operated into the 1920s. Milling machine production included profiling machines (Fig.1), introduced in 1915, and continuous millers (Fig.2), offered in 1920. The continuous millers were available with three or four spindles. The profiling machines were designed for the rapidly expanding metalworking firms which were being inundated with European munitions orders. Betts noted that they had just sold (in 1916) 120 of the profiling machines to one manufacturer. *(illustrations on next page)*

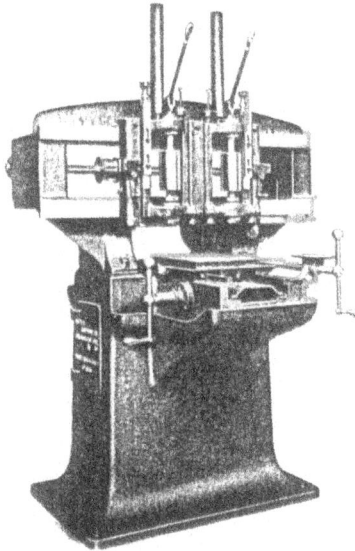

Fig.1

Fig.2

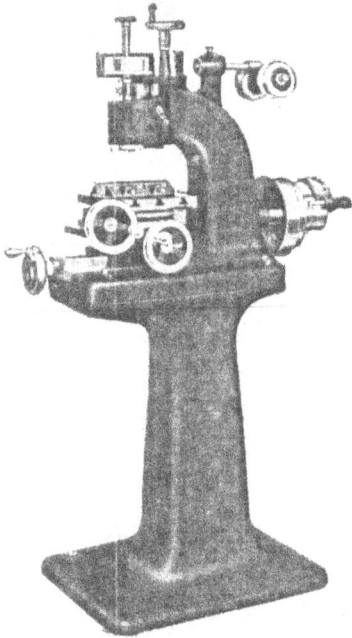

BICKETT MACHINE & MFG. CO

Cincinnati, OH

Formed about 1915 by C.A. Bickett to make the Bickett No.0 vertical milling and profiling machine. The machine was available as a bench machine or with a matching stand *as shown at left.* In 1923, Bickett reorganized as the BICKETT MILLER CO.

BICKFORD MACHINE CO., Greenfield, MA

Formed in 1910 by Oramel S. Bickford as a reorganization of BICKFORD & WASHBURN. In 1917, Bickford sold the company to the Greenfield Tap & Die Co. which operated it for a time as the Bickford Machine Co. division of Greenfield Tap & Die Co. Bickford remained as manager until 1920. Production included small milling machines previously made by Bickford & Washburn. The machines were furnished with both hand and power feed and equipped with vertical milling attachments.

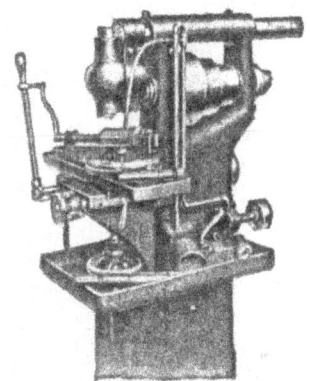

A Medium Weight Miller with both Power and Hand Feed

Moreover, the vertical milling attachment for this machine puts it in an exclusive class. This attachment is strong and self-contained, and can be applied with ease.

There are other points about this machine deserving of your attention. **Write for the bulletin.**

Bickford Machine Co.
Greenfield, Mass.

BICKFORD & WASHBURN, Greenfield, MA

A partnership of Oramel S. Bickford and H.L. Washburn, formed in 1908 to make thread milling machines. In 1910, the firm introduced a No.1 milling machine, *shown at right*, with hand or power feed, 15" table travel, 3" cross travel, and 7" vertical travel. The partnership broke up in 1910; Bickford then reorganized as the BICKFORD MACHINE CO. that continued production of the No.1 milling machine.

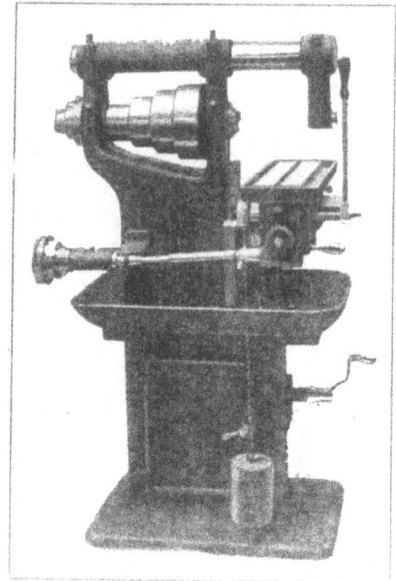

BILLINGS & SPENCER CO., Hartford, CT

A partnership of Charles E. Billings (1835-1920) and Christopher Spencer (1833-1922) formed in 1869 as the Roper Sporting Arms Co. and reorganized as Billings & Spencer Co. in 1873. The company's primary products were drop forgings of all types, especially for arms makers and sewing machine makers. A line of drop hammers was offered to other manufacturers. Production of milling machines began in 1908 with the introduction of an "under" die trimmer (Fig.1). In this type of milling machine the cutter is located below the workpiece. The line expanded in 1920 when a B.M.W. Hanson-designed die sinking machine (Fig.2) was introduced. The machines were a natural outgrowth of the company's forging and drop hammer business. The PRATT & WHITNEY CO. also introduced a die sinking machine in 1920, also designed by its employee B.M.W. Hanson. The Billings & Spencer machine must have been made under license from Pratt & Whitney.

Fig.1

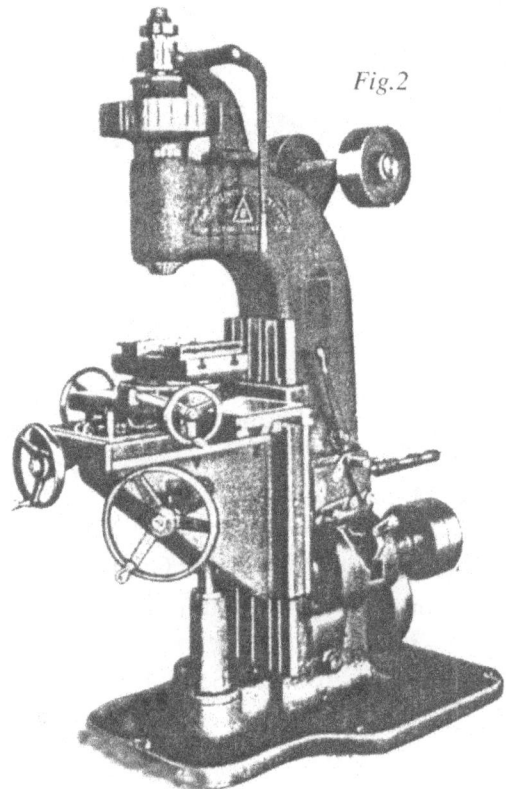

Fig.2

BLAKE & JOHNSON, Waterbury, CT

A partnership of James P. Blake and Charles W. Johnson, founded in 1852 to make small rolling and bending mills for jewelry and silver goods manufacturers. Milling machines were later offered for milling dies used by their forming machinery. Production included vertical milling machines (Fig.1) designed by F.B. Manville and introduced in 1892. A much heavier machine (Fig.2) was brought out in 1898, and a new vertical milling machine, with rotary table (Fig.3), was offered in 1901. The machine was built in four sizes, including a No.5 with 42" table travel and 14" cross feed.

Fig.1

A New Vertical Milling Machine.

Fig.2

36"×36" Vertical Spindle Milling Machine

Fig.3

No. 5 Vertical Milling Machine.

BLISS CO., E.W., Brooklyn, NY

Founded in 1881 by Eliphalet W. Bliss (1836-1903) to make a line of presses and other metal-forming machinery. For a few years, beginning in 1881 and through about 1895, Bliss also offered metal cutting machine tools. Milling machine production began in 1883 with the introduction of a vertical machine (Fig.1) that had been developed for its own use. By 1887 production included vertical machines offered in two sizes (Figs.2-3) and two-spindle machines offered in only one size (Figs.4-5). The two-spindle machine was offered, in improved form, (Fig.6) as late as 1895. *(illustrations on next two pages)*

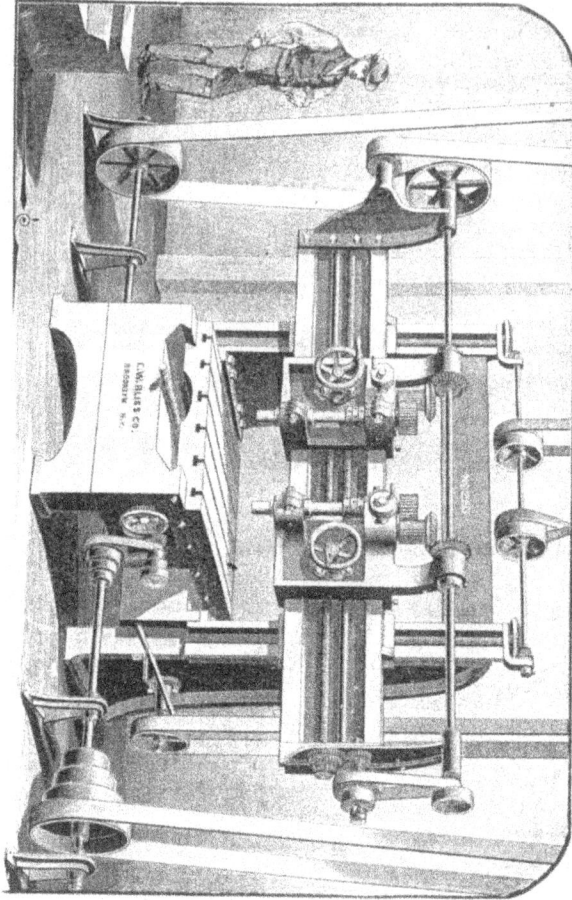

Fig. 4

Fig. 5

30

Price, ...		$2,200
Weight, about, ...		13,000 lbs.
Length of table as desired, or . . .		48 ins.
Width of table, ...		30 ins.
Width between housings, . . .		54 ins.
Least width between centres of spindles, . .		8 ins.
Vertical adjustment of cross head, . .		24 ins.
Cone faces, 16, 14, 12 and 10 ins. dia × 3¼ in. wide.		
Floor space over all, F. & B. × R. & L, . .		5 × 12 ft.
Height over all, ...		9 ft.

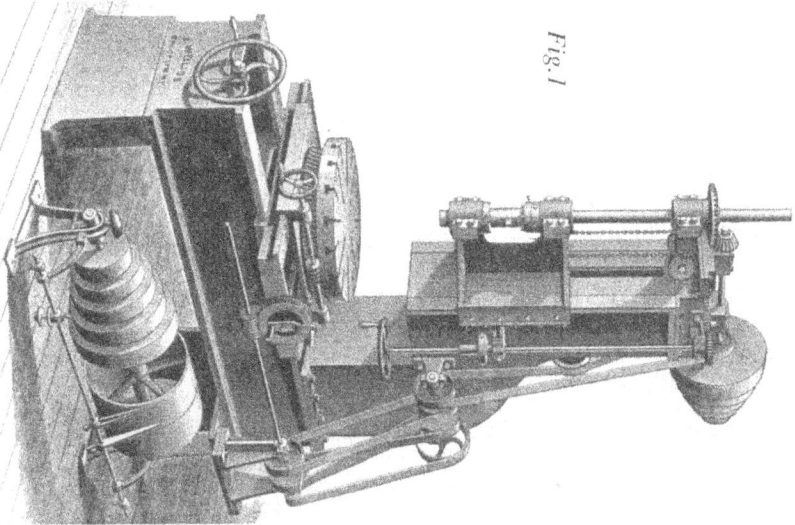

Fig. 1

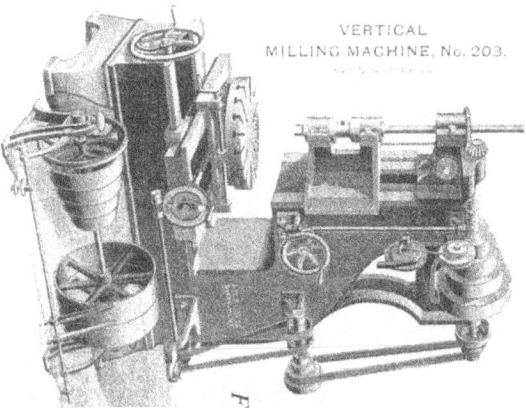

VERTICAL
MILLING MACHINE, No. 203.

Fig. 2

Fig. 3

	No. 202 16 inch	No. 203 24 inch
Price,	$475.00	$1,100.00
Weight,	3,800 lbs.	5,700 lbs.
Length of head,	18 ins.	24 ins.
Adjustment of head,	16 ins.	15 ins.
Distance back from centre of spindle,	18 ins.	24 ins.
Standard cut back,	8¼ ins.	14 ins.
Width of standard,	16 ins.	17 ins.
Diameter of table,	20 ins.	24 ins.
Length of longitudinal feed,	16 ins.	24 ins.
Length of cross feed,	16 ins.	20 ins.
Distance table to top of throat,	8 ins.	8 ins.
Distance bed to top of throat,	19 ins.	20 ins.
Floor space over all, front to back,	6 ft. 9 ins.	6 ft. 6 ins.
Floor space over all, right to left,	3 ft. 5 ins.	4 ft. 4 ins.
Height to top of cone,	8 ft. 8 ins.	8 ft. 9 ins.
Height to top of spindle, when up,	8 ft. 8 ins.	9 ft. 5 ins.

TWO-SPINDLE MILLING MACHINES
Nos. 220 and 221.

Fig.6

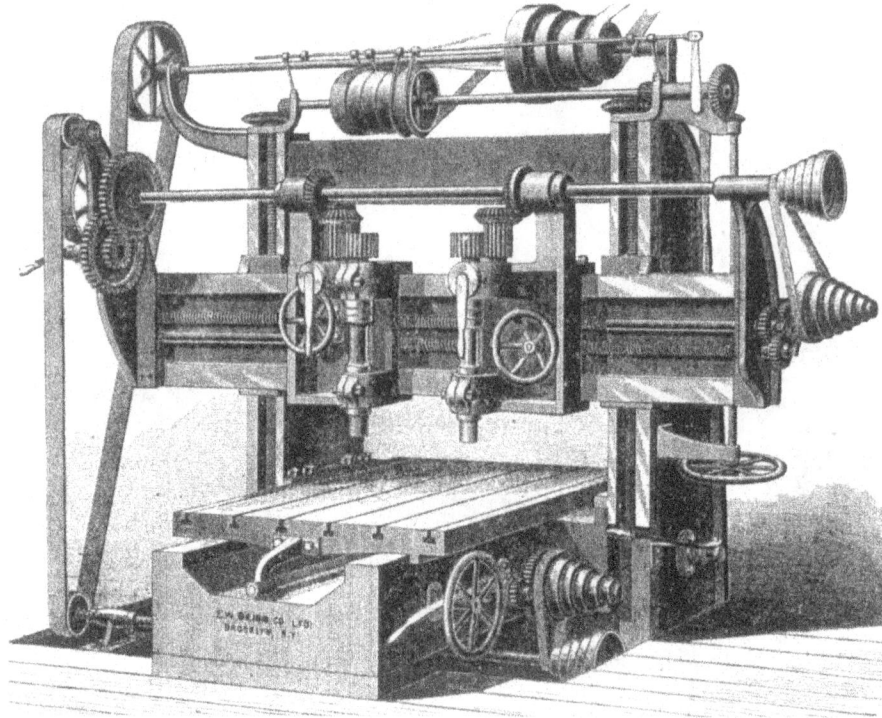

BOGERT, JOHN L., Flushing, L.I., NY

Bogert, who had been a partner in Grant & Bogert, began operating under his own name in 1884 when the partnership was dissolved. He remained active until at least 1906. Products offered in 1893 included turret machines, lathes, and milling machines. The milling machines were of the Duplex double-column type *as shown below*. Table feed was 25" or an optional 30".

BOND FOUNDRY & MACHINE CO., Manheim, PA

Maker, in 1916, of the BOWEN milling machine designed by James E. Bowen, primarily for thread milling operations. The head could be rotated so the machine could be used as either a vertical or horizontal milling machine.

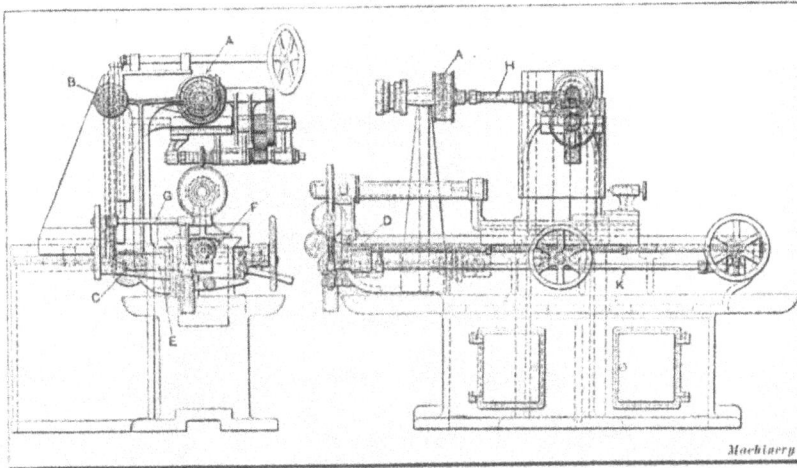

BRAINARD MILLING MACHINE CO., Hyde Park, MA

Formed in 1871 by Amos Brainard (1824-1905) as a reorganization of the UNION VISE CO., which had made milling machines of his design. New Brainard designs were based on his patents of January 17, 1871, January 31, 1871, October 1, 1872, August 9, 1887, February 28, 1888, and April 24, 1888. 1871 production included universal milling machines (Fig.1), patented October 1, 1872, and priced at $200. They were offered with dividing heads (Fig.2), patented January 17, and January 31, 1871. By 1874, Brainard was advertising "milling machines of all sizes and styles, weight from 600 pounds to 10,000 pounds". These included a new No.3 universal mill (Fig.3) with 18" table travel, 5" cross travel and 16" vertical travel, weighing 1,700 pounds. *(continued next eight pages)*

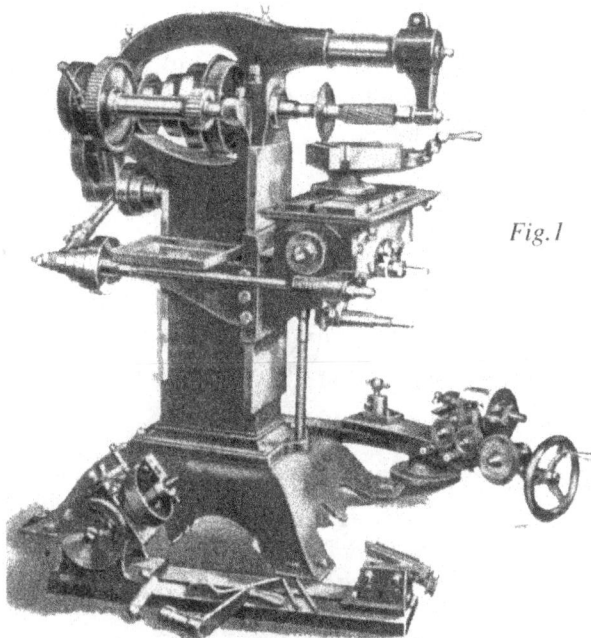

Fig.1

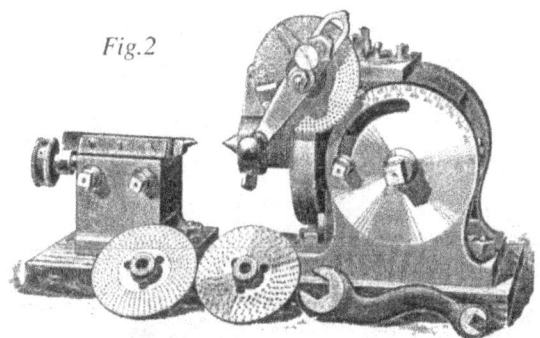

Fig.2

Fig.3

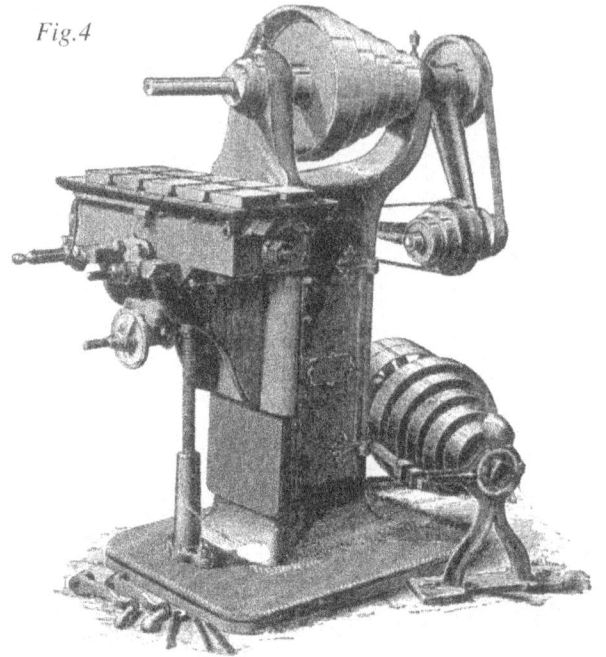

Fig.4

1880 production included smaller column milling machines without overarms (Fig.4); plain milling machines (Fig.5) with 24" x 6" tables, weighing 950 pounds; and index milling machines (Fig.6) weighing only 600 pounds. By 1884, the line had expanded to include Standard Universal milling machines in No.1 size (Fig.7) with 36" table travel, 8" cross travel and 24" vertical travel, priced at $850; No.2 size (Fig.8) with 24" table travel, 7" cross travel and 21" vertical travel, priced at $650; No.3 size (Fig.9) with 20" table travel, 5" cross travel and 18" vertical travel, priced at $500; No.4 size (Fig.10) with 18" table travel, 5" cross travel and 16" vertical travel, priced at $400; No.5 size (Fig.11) with 9" table travel, 3" cross travel and 7" vertical travel, priced at $250. Also available were large hand milling machines (Fig.12), priced at $250 and small hand milling machines priced at $175; No.7 plain millers (Fig.13) weighing 1,400 pounds, priced at $310; and No.8 plain millers (Fig.14) weighing 950 pounds and priced at $275. Attachments included a universal head with back center (Fig.15), priced at $150 and hand spiral cutting device (Fig.16), also priced at $150.

Fig.5

Fig.6

No. 1 STANDARD UNIVERSAL
MILLING MACHINE.

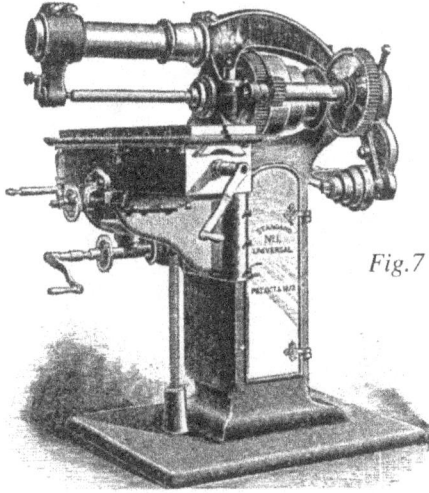

Fig.7

No. 2 STANDARD UNIVERSAL
MILLING MACHINE.

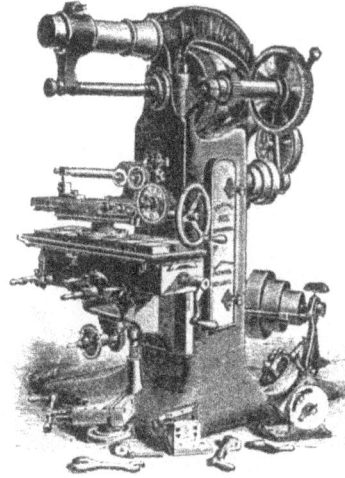

Fig.8

No. 3 STANDARD UNIVERSAL
MILLING MACHINE.

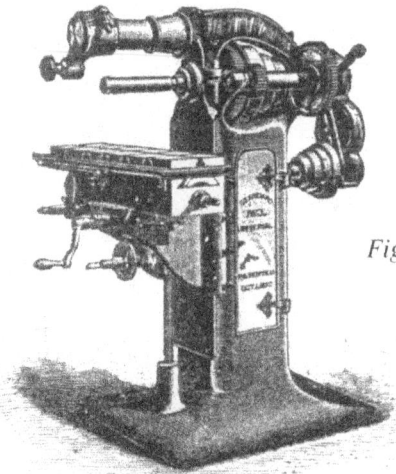

Fig.9

No. 4 STANDARD UNIVERSAL
MILLING MACHINE.

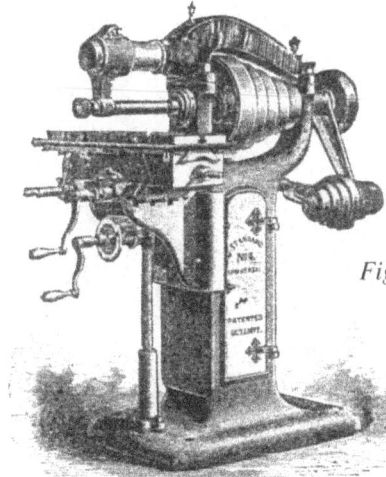

Fig.10

No. 5 STANDARD UNIVERSAL
MILLING MACHINE.

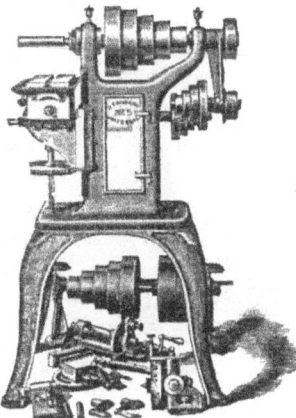

Fig.11

LARGE HAND MILLING MACHINE.

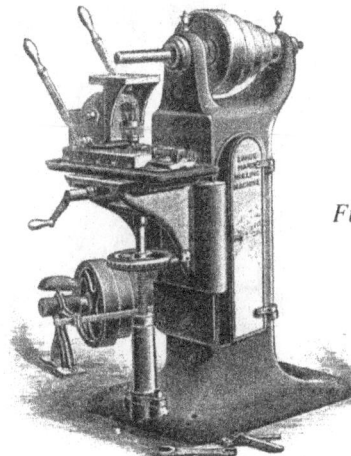

Fig.12

No. 7 PLAIN MILLING MACHINE.

Fig.13

No. 8 PLAIN MILLING MACHINE

Fig.14

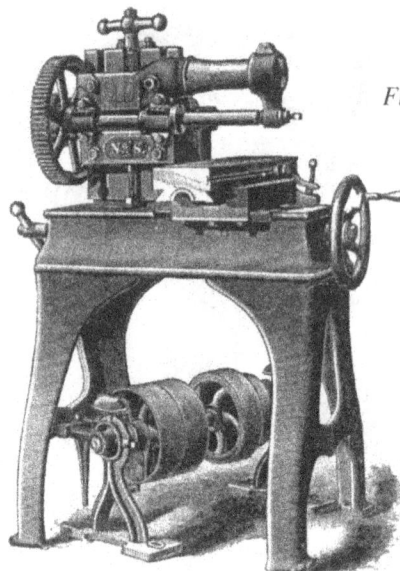

UNIVERSAL HEAD WITH BACK CENTER.

Fig.15

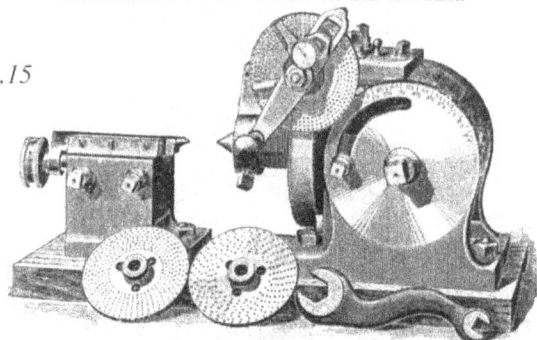

HAND SPIRAL CUTTING DEVICE.

Fig.16

Fig.17

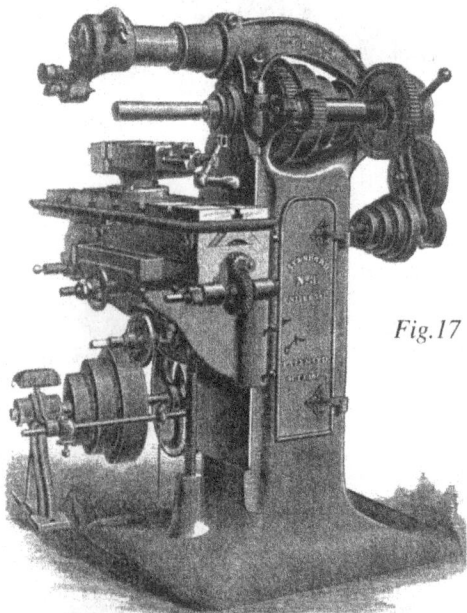

Improved Brainard Universal Milling Machine

Fig.18

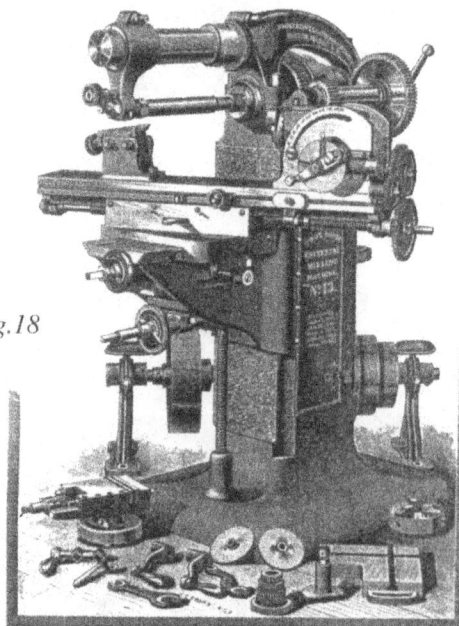

Brainard Universal Milling Machine

A new line of universal milling machines, patented August 9, 1887, February 28, 1888, and April 24, 1888, was brought out in 1888 (Fig.17), with further improvements in 1889 (Fig.18). Cam cutting attachments (Fig.19) were also offered in 1889. 1893 production included Standard Universal milling machines in No.1 size (Fig.20), No.2 size (Fig.21), No.3 size (Fig.22), No.4 size (Fig.23), and No.5 size (Fig.24) as detailed in Fig.25; tool-room universal milling machines in No.11, No.12, No.13, No.14 (Fig.26), and No.15 (Fig.27) sizes; plain milling machines in No.20 (Fig.28), No.21, No.22 (Fig.29) and No.24 sizes. Attachments included a new gear cutting device (Fig.30), new universal head (Fig.31), combination head and patent back center (Fig.32), cam cutting attachment (Fig.33), spiral cutting attachment (Fig.34), rack cutting attachment (Fig.35), vertical milling attachment (Fig.36), and rotary milling attachment with overarm brace (Fig.37). Other milling machines included the No.6 (Fig.38), No.7 (Fig.39), and No.8 plain; long feed horizontal machines in 50" (Fig.40), 72" (Fig.41) and 84" (Fig.42) sizes; small index mills the same as those offered in 1880, and automatic cam cutting machines (Fig.43). Brainard merged with the JOHN BECKER MFG. CO. in 1899 to form the BECKER-BRAINARD MILLING MACHINE CO. where he was president until retiring in 1901.

Fig.19

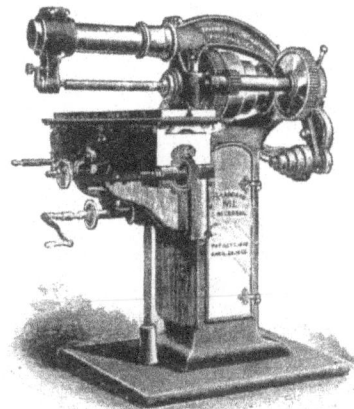

No. I Universal Milling Machine.

Fig.20

Fig.21

No. 2 Universal Milling Machine.

No. 3 Milling Machine

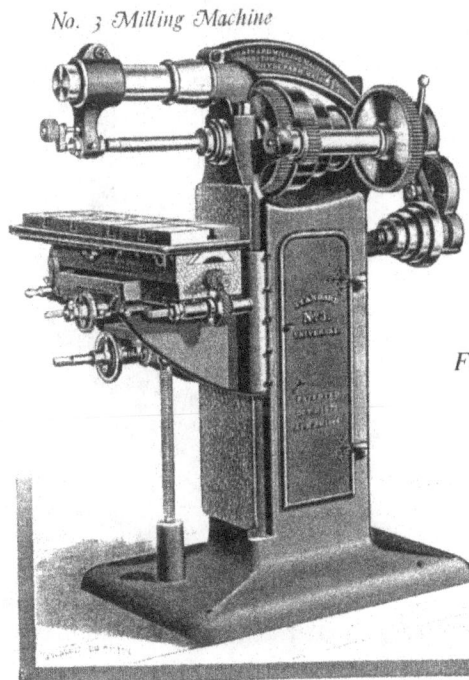

Fig.22

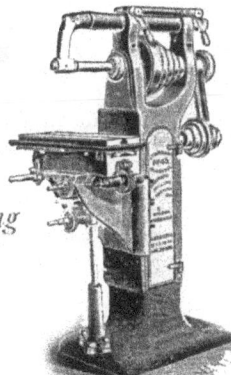

Fig.23

No. 4½ Standard Universal Milling Machine.

With New Patent Quick Return Movement.

No. 5 Standard Universal Milling Machine.

COMPLETE.

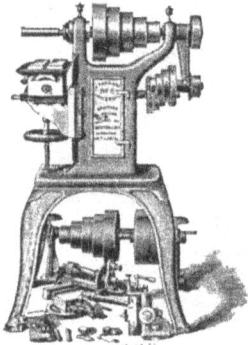

Fig.24

Fig.25

Brainard's Standard Universal Milling Machines.

LIST OF SIZES, DETAILS, AND PRICES.

	No. 1.	No. 2.	No. 3.	No. 4.	No. 4½.	No. 5.
Automatic longitudinal feed	36 in.	26 in.	22 in.	18 in.	12 in.	9 in.
Vertical range below spindle	24 "	21 "	18 "	16 "	15 "	7 "
Cross motion in line with spindle . . .	10 "	8½ "	6½ "	5 "	5 "	3 "
Number of speeds for driving belt . . .	6	6	6	5	5	5
Width of driving belt	4 in.	3½ "	2¾ "	3½ in.	2 in.	2 in.
Number of changes for feed	4	4	4	4	4	4
Size of cutter that can be used under arm	12 in.	10 in.	8 in.	8 in.	No arm.	
Will mill out from column	18 "	16 "	13 "	12 "	No arm.	
Weight, complete, boxed for shipment .	4,800 lbs.	3,500 lbs.	2,350 lbs.	2,000 lbs.	1,200 lbs.	900 lbs.
Price, with countershaft, no attachments						
Price, complete, with regular attachments						
Speed of countershaft	100 rev.	120 rev.	120 rev.	120 rev.	130 rev.	130 rev.
Diameter tight and loose pulleys . . .	18 in.	16 in.	14 in.	14 in.	12 in.	10 in.
Width tight and loose pulleys	4½ "	4 "	3½ "	3½ "	2½ "	2½ "

Nos. 1, 2, and 3 are the only sizes made with Back Gears. Nos. 1, 2, 3, and 4 have the overhanging arm to support outer end of cutter arbor. Nos. 4½ and 5 do not have this arm. except as it is added to No. 4½ as an extra.

Brainard's No. 13
Tool-Room Universal Milling Machine.

Fig.26

Brainard Tool Room Milling Machine No. 15.

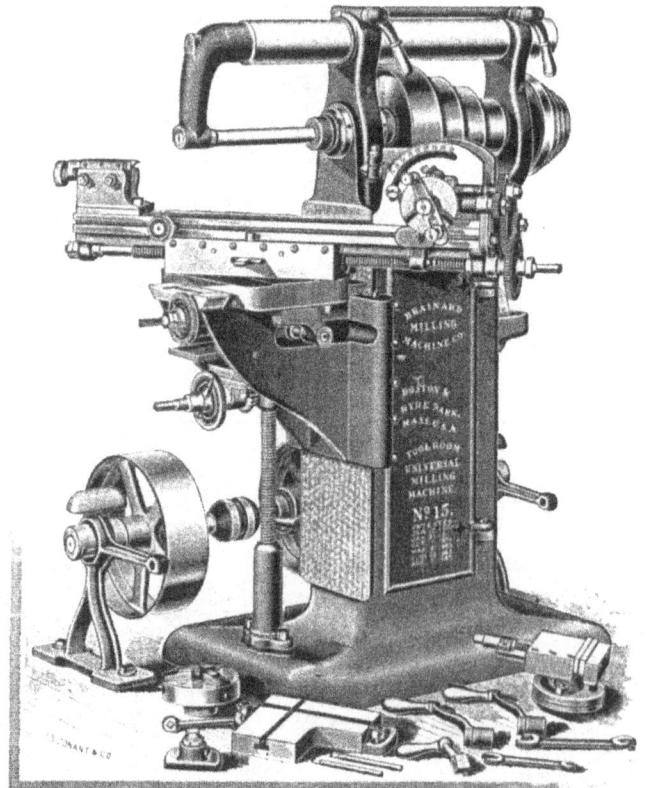

Fig.27

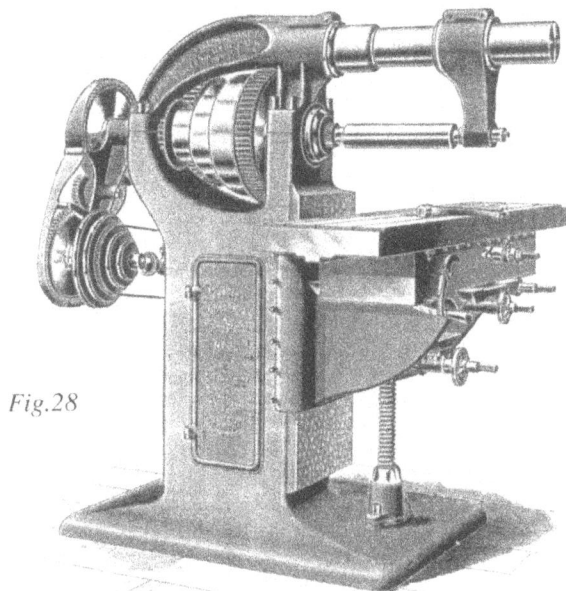

Upright Plain Milling Machine, No. 20.

Fig.28

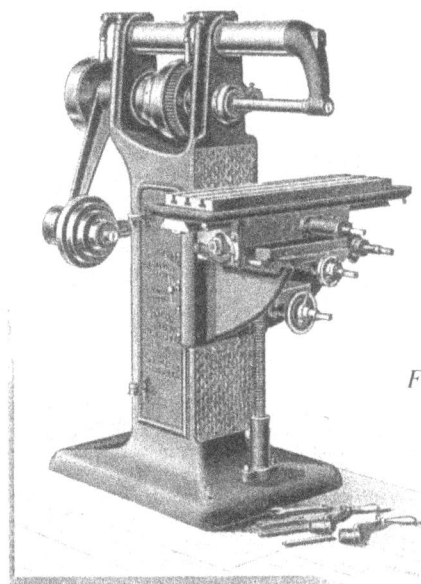

Upright Plain Milling Machine.
SCREW FEED, RACK AND PINION RETURN

Fig.29

Fig.30

Brainard's New Patent Universal Head.

Fig.31

Brainard's
New Combination Head and Patent Back Centre.

Fig.32

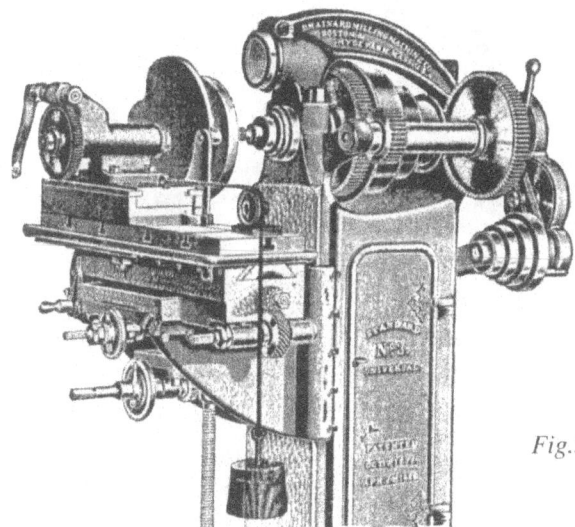

Fig.33

Cam Cutting Attachment.

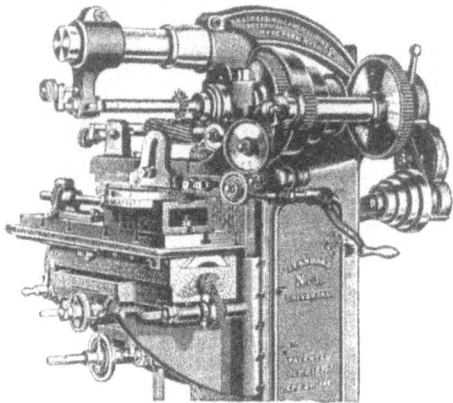

Automatic Spiral Cutting Attachment.

Fig.34

Rack-Cutting Attachment.

Fig.35

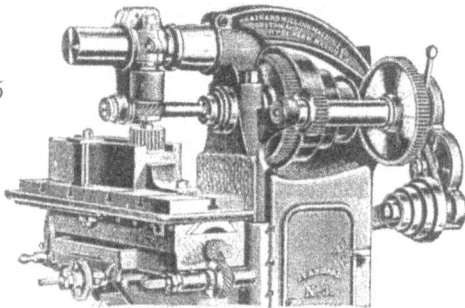

Vertical Milling Attachment.

Fig.36

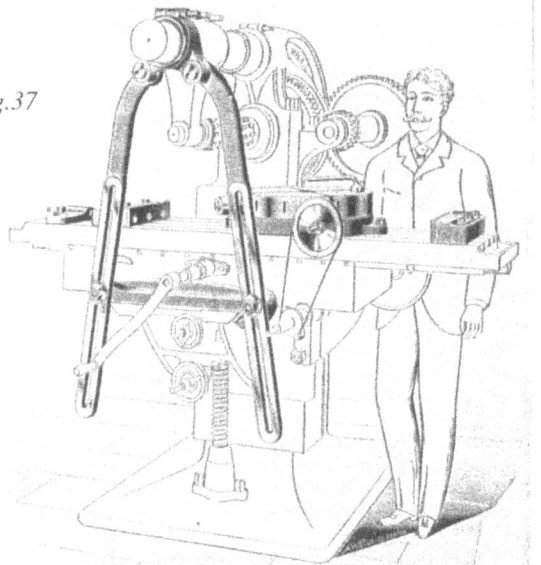

Fig.37

Cut of Table Vise Holding Work Whole Length of Table, also Cut of Automatic Rotary Milling Attachment, and of "Harness."

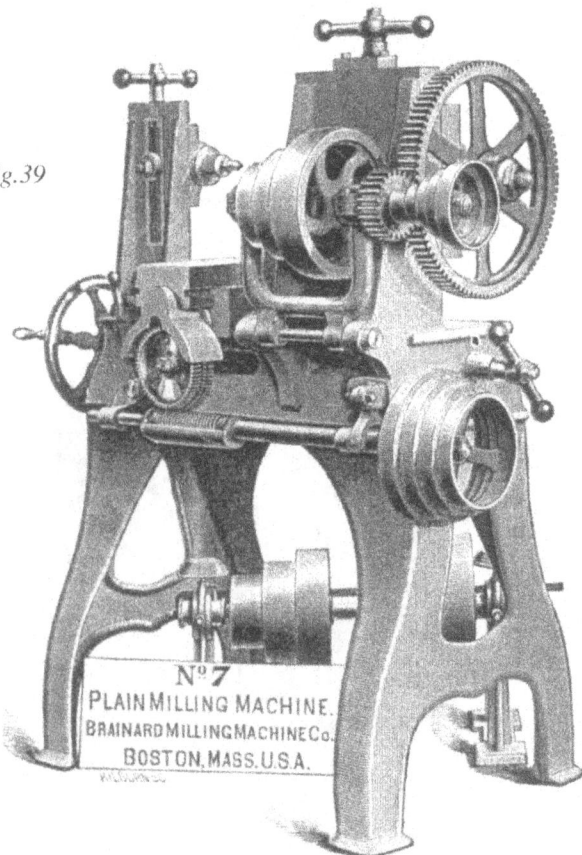

Fig.39

No 7
PLAIN MILLING MACHINE.
BRAINARD MILLING MACHINE CO.
BOSTON, MASS. U.S.A.

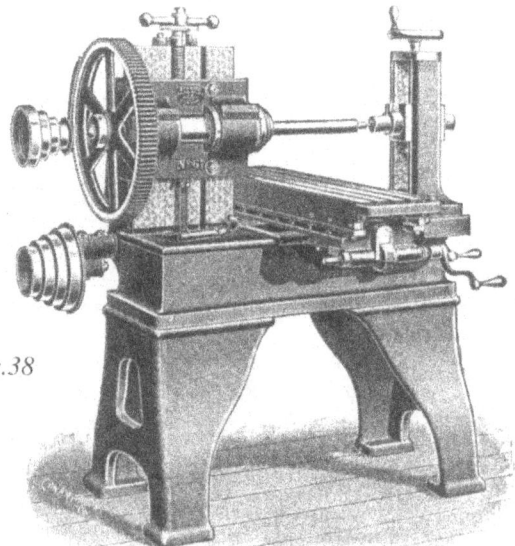

Fig.38

Brainard Lincoln Pattern Milling Machine.

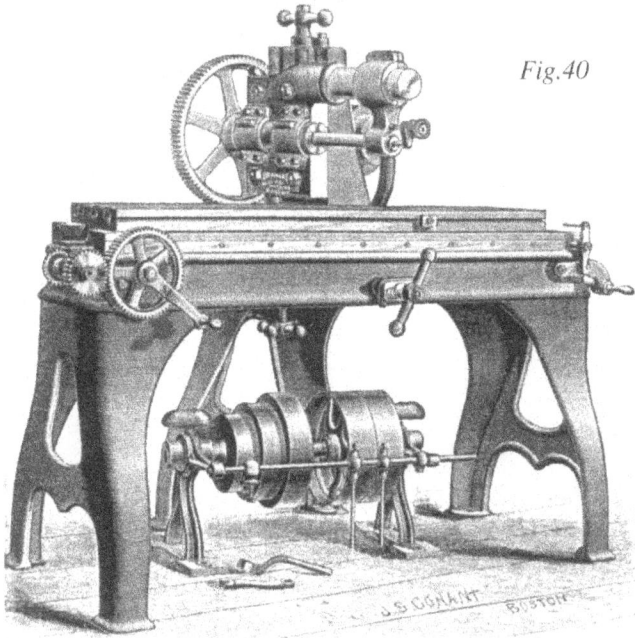

Fig.40

Fig.41

Fig.42

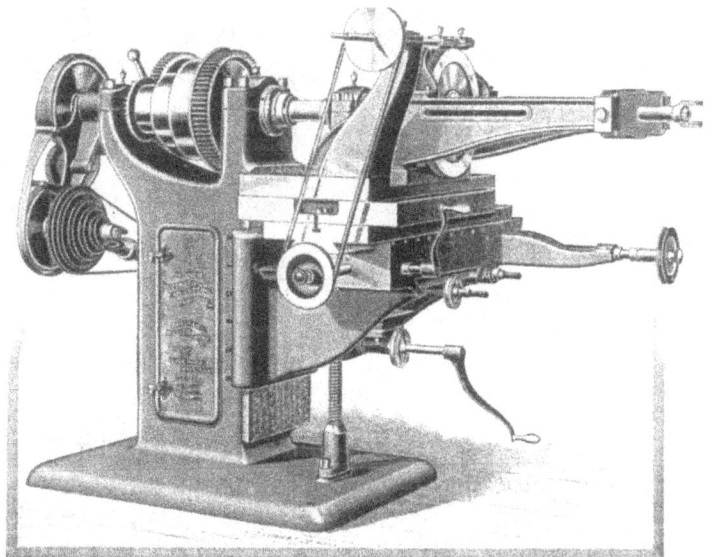

Fig.43

BRISTOL MACHINE TOOL CO., Bristol, CT
See C.G. GARRIGUS MACHINE CO.

BROMBACHER'S SONS, JACOB, New York, NY

Maker, in 1894, of a heavy vertical milling machine equipped with a rotary table and with 30" table motion, 23" cross feed and 23" vertical motion. Note the rope drive for the spindle and the idler pulley which maintains tension as the spindle is moved up and down.

VERTICAL MILLING MACHINE

BROWN & SHARPE, J.R., Providence, RI,

later **BROWN & SHARPE MFG. CO.**, Providence, RI

A partnership of Joseph R. Brown (1810-1876) and his former apprentice Lucian Sharpe (1830-1899) formed March 1, 1853. Early products included measuring tools, clocks and sewing machines. Machine tool production began in 1861 when Brown designed a screw machine for firearms makers. Milling machine production began in 1862 when Brown was approached by Frederick W. Howe, superintendent of the PROVIDENCE TOOL CO., for help in designing a machine for cutting the grooves in twist drills. The result was the first universal milling machine (Fig.1) which was produced, with many enhancements, into the 1970s. On January 1, 1868, Frederick W. Howe and Thomas McFarlane were admitted to partnership; the firm reorganized as the Brown & Sharpe Mfg. Co. and continues under that name today. Its 1868 catalog offered the Howe-designed plain milling machine (Fig.2) previously made by the Providence Tool Co. In 1872 the firm employed 300 workmen. By 1900 employment had grown to 2000 and factory floor space to seven acres. Machine tool production ceased in 1991. In 1876 the firm introduced a larger version of the universal milling machine (Fig.3) with 22" table travel and 6½" cross travel. Equipped with an overarm, it was priced at $1,100 and later designated the No.3 universal miller. Production of the original No.1 universal miller, priced at $700, continued with only minor changes as shown in an 1882 article (Fig.4). 1881 production included new No.1 plain millers (Fig.5) with 9" table travel, priced at $260; and No.2 plain millers (Fig.6) with 18" table travel, priced at $340. A large plain milling machine (Fig.7) with 48" table travel and 7¼" cross travel, was introduced in 1883 and later designated the No.6 plain miller. *(continued on next seventeen pages)*

Fig.1

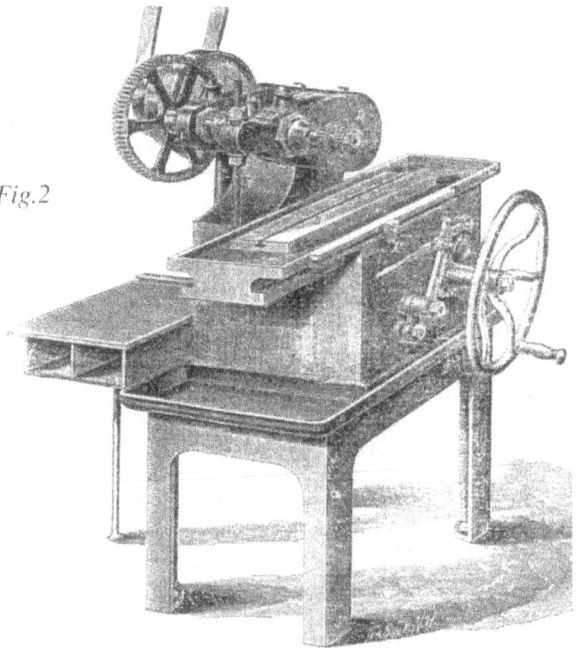

Fig.2

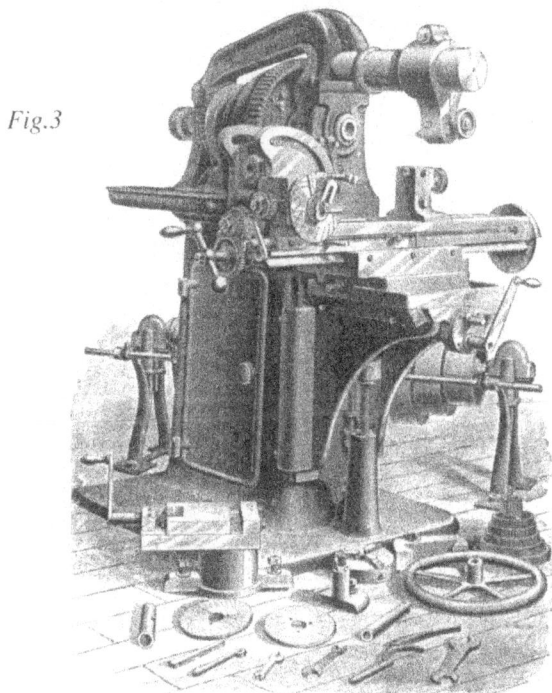

Fig.3

Fig.4

SMALL UNIVERSAL MILLING MACHINE.

Fig.5

Small Plain Milling Machine.

Fig.6

Plain Milling Machine.

Fig.7

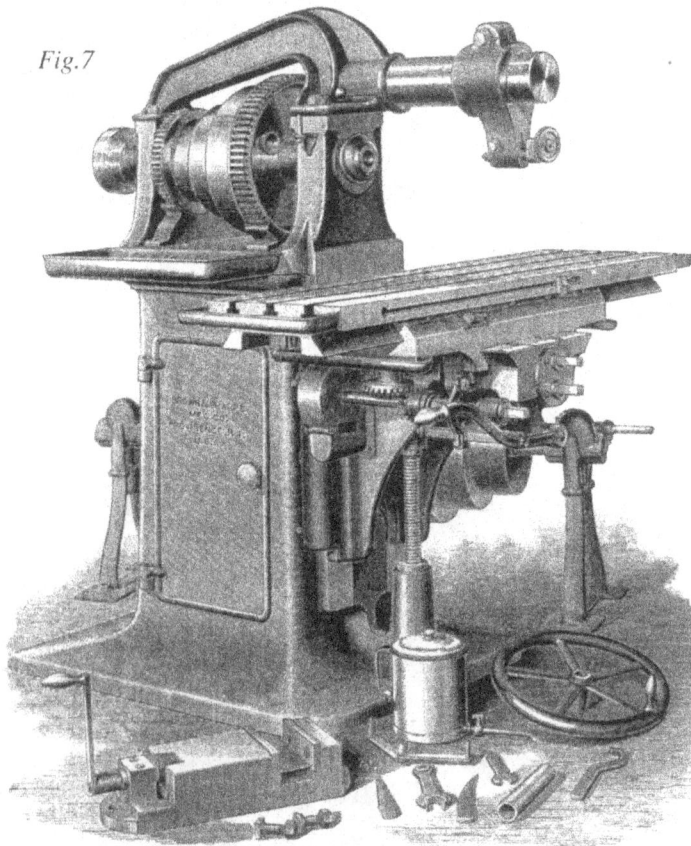

New Large Plain Milling Machine

Fig.8

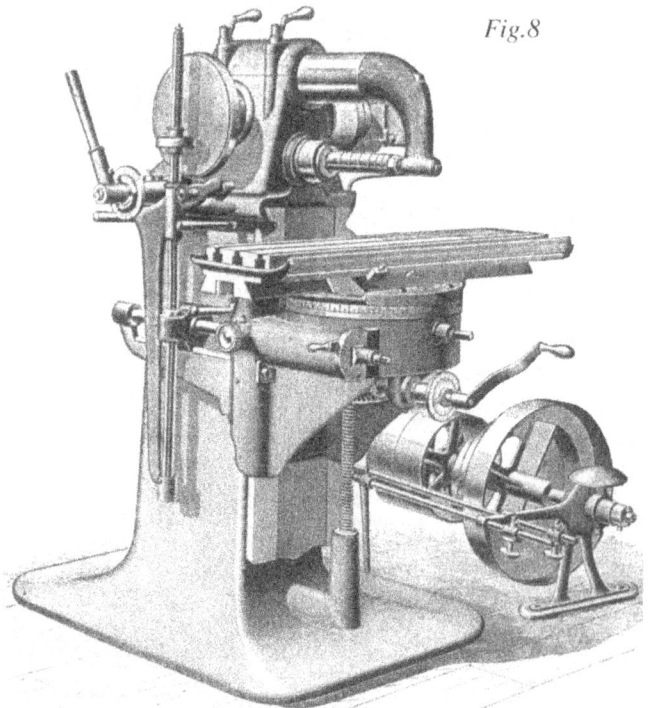

In 1885, the C.E. LIPE line of milling machines, along with the patents, was purchased and introduced as the Brown & Sharpe No.4 heavy universal miller (Fig.8) with 36" table travel; No.5 plain milling machine (Fig.9) with 51" table travel; No.7 plain miller with 60" table travel; and Lipe's patent dividing head (Fig.10). The No.3 plain miller (Fig.11) with 12" table travel, 4½" cross travel and weighing 2,550 pounds was also introduced in 1885. The No.2 vertical milling machine (Fig.12), the first vertical, was introduced about 1887. Weighing 3,300 pounds, it had 36" table travel and 12" cross travel. The first major change to the No.1 universal milling machine, the addition of an overarm (Fig.13) was made in 1889. A new series of milling machines was introduced at the Columbian Exposition in 1892. The series included the No.8 plain (Fig.14) with a 66" x 16" table and weighing 5,000 pounds; a plain miller (Fig.15) with 60" x 14" table and weighing 3,435 pounds; a plain miller (Fig.16) with 40" x 10" table and weighing 1,645 pounds; and No.2 universal (Fig.17) with 35" x 8" table, weighing 1,750 pounds. New design No.4 (Fig.18) plain millers with 42" table travel and 8" cross travel, weighing 3,800 pounds, and No.5 plain millers with 46" table travel and 9¾" cross travel, weighing 5,500 pounds, were introduced in 1898. Improved universal milling machines, the "1900 design", were introduced in 1900 and included No.2, No.3 (Fig.19), No.4 and No.5 sizes.

Fig.9

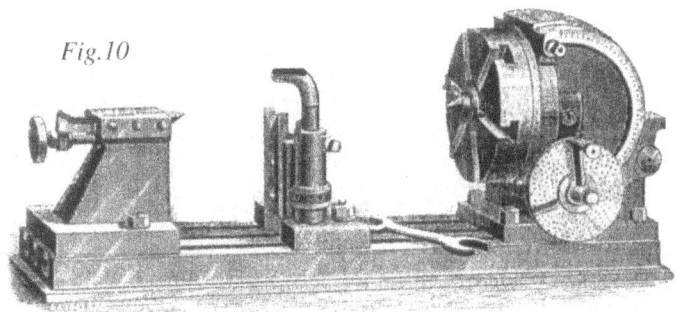

Fig.10

Fig.11

Brown & Sharpe's No. 3 Plain Milling Machine.

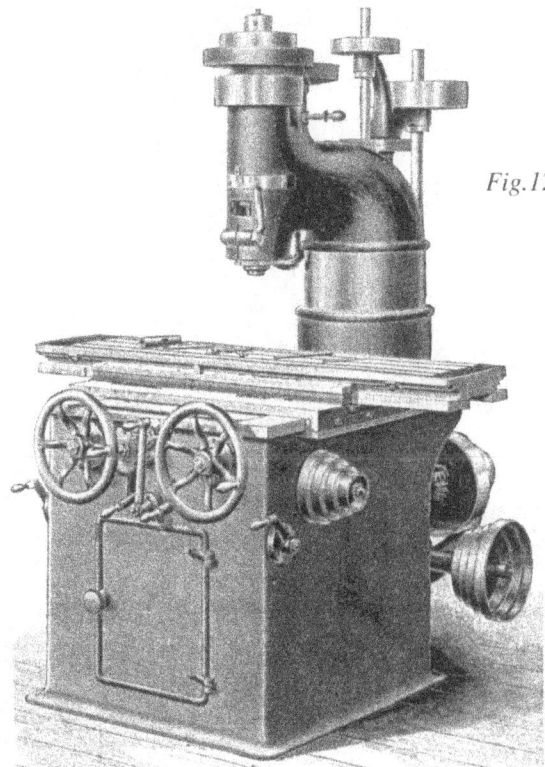

Fig.12

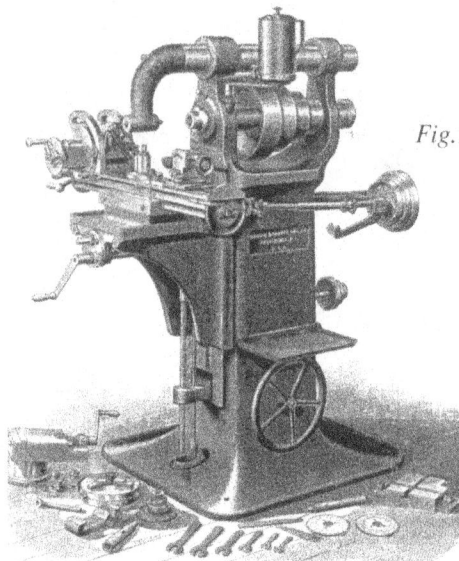

Fig.13

No. 1 Milling Machine with Overhanging Arm.

Fig.14

Brown & Sharpe No. 8 Plain Milling Machine.

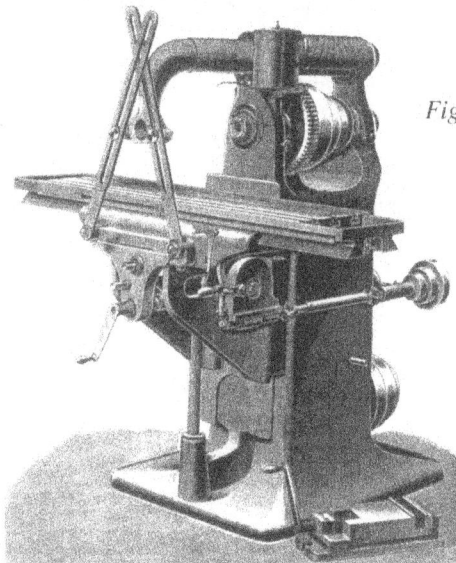

Fig.15

Fig.16

Plain Milling Machine.

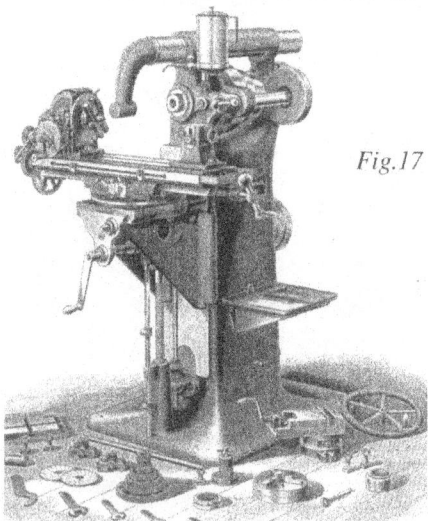

Fig.17

Universal Milling Machine.

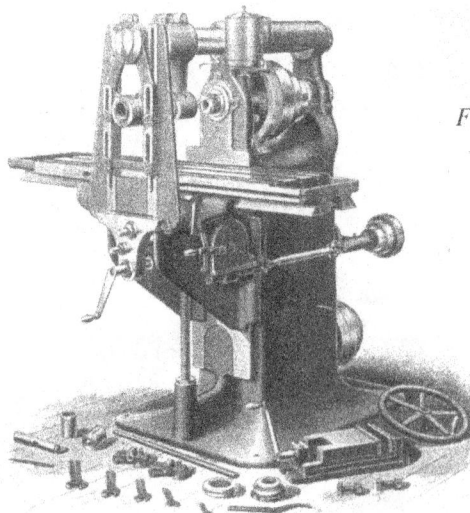

Fig.18

No. 4.

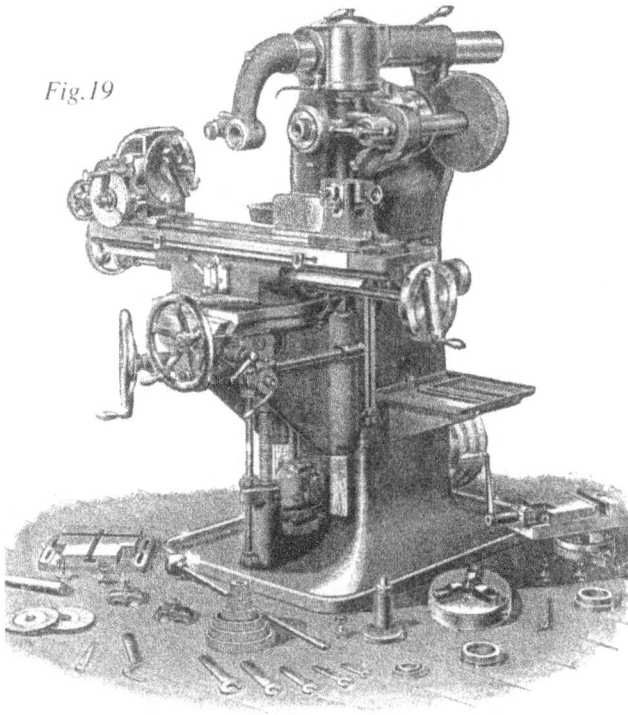

Fig.19

NO. 3

No. 1
20 in. x 6 in. x 18 in.

UNIVERSAL MILLING MACHINE.

Patented Feb. 5, 1884; Feb. 14, May 23, 1893;
Aug. 29, 1899.

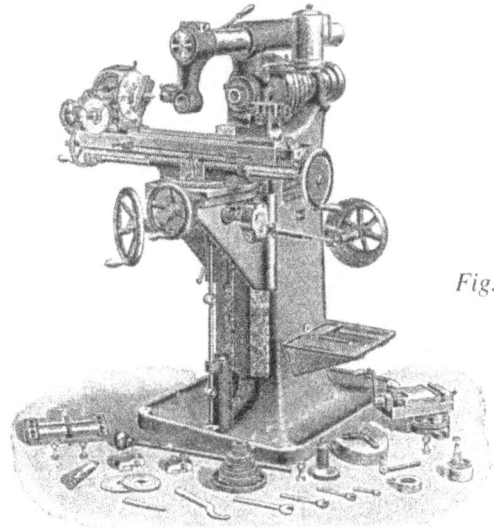

Fig.20

Brown & Sharpe's 1902 catalog listings included universal milling machines in No.1 (Fig.20), No.2 (Fig.21), No.3 (Fig.22), and No.4 (Fig.23) sizes; plain milling machines in No.0 (Fig.24), No.1 (Fig.25), No.2 (Fig.26), No.3 (Fig.27), No.4 (Fig.28), No.5 (Fig.29), No.12 (Fig.30), No.13 (Fig.31), and No.24 (Fig.32) sizes; vertical milling machines in No.2 (Fig.33) and No.5 (Fig.34) sizes; high speed milling attachments (Fig.35); vertical spindle milling attachments (Fig.36); slotting attachments (Fig.37); universal milling attachments (Fig.38); rack cutting attachments (Fig.39); and index centers (Fig.40).

No. 2
23 in. x 6 1-2 in. x 17 1-2 in.

UNIVERSAL MILLING MACHINE.
With Hand or Power Vertical Feed.

Patented Feb. 5, 1884; Feb. 14, 1893; May 23, 1893;
Aug. 29, 1899

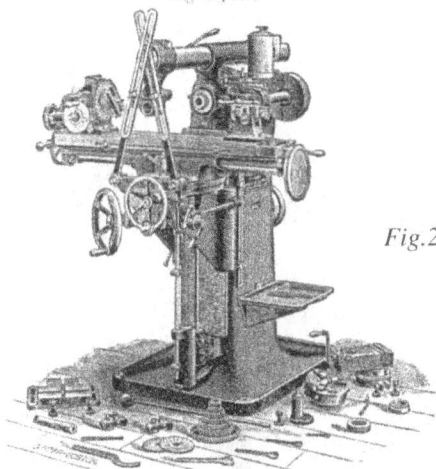

Fig.21

No. 3
28 in. x 7 1-2 in. x 18 1-2 in.

UNIVERSAL MILLING MACHINE.

Patented Feb. 5, 1884; Feb. 14, May 23, 1893;
Aug. 29, 1899.

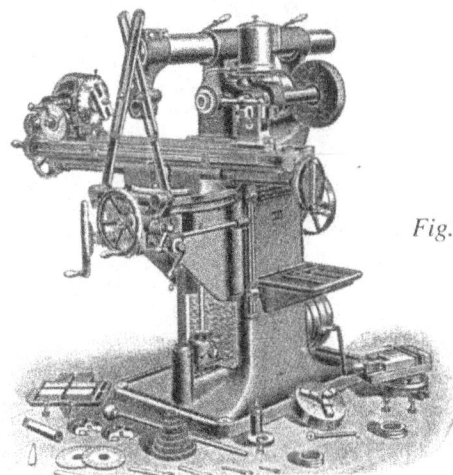

Fig.22

46

No. 4
33 1-2 in. x 8 1-2 in. x 19 in.

UNIVERSAL MILLING MACHINE.

Patented Feb. 5, 1884; Feb. 14, 1887;
May 23, 1893; Aug. 29, 1899.

Fig. 23

No. 0
16 in. x 4 1-4 in. x 14 1-2 in.

PLAIN MILLING MACHINE.

Screw Feed Machine Patented May 23, 1893; Aug. 29, 1899.

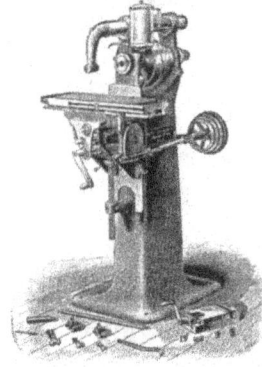

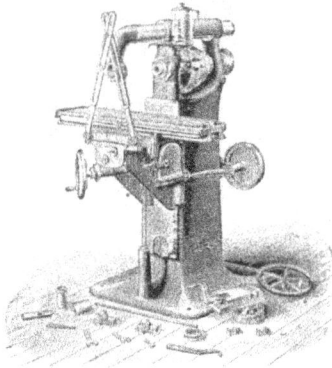

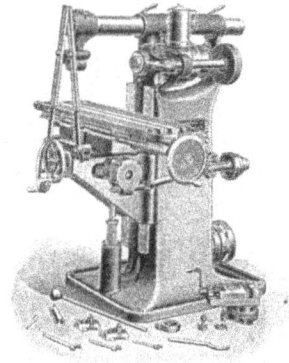

Fig. 24

Cut Shows Rack Feed Machine.

No. 1
24 in. x 6 1-2 in. x 18 1-2 in.

PLAIN MILLING MACHINE.

Screw Feed Machine Patented May 23, 1893; Aug. 29, 1899.

Fig. 25

Cut Shows Rack Feed Machine.

No. 2
28 in. x 6 1-2 in. x 18 1-2 in.

PLAIN MILLING MACHINE.

Screw Feed Machine Patented
May 23, 1893; Aug 29, 1899.

Fig. 26

CUT SHOWS SCREW FEED MACHINE.

No. 3
34 in. x 8 in. x 19 3-4 in.

PLAIN MILLING MACHINE.

Screw Feed Machine Patented Feb. 6, 1900.
Rack Feed Machine Patented Jan. 18, 1898; Feb. 6, 1900.

Fig. 27

Cut Shows Rack Feed Machine.

No. 4
42 in. x 8 3-4 in. x 20 in.

PLAIN MILLING MACHINE.

Patented Oct. 18, 1892; Jan. 18, 1898;
Feb. 6, 1900.

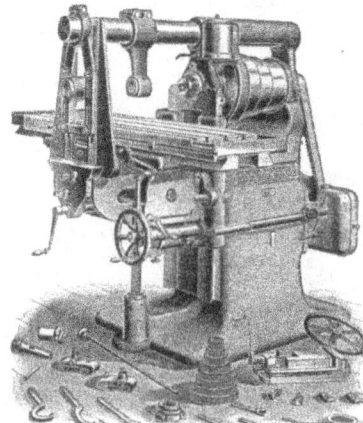

Fig. 28

No. 5

48 in. x 9 3-4 in. x 19 1-2 in.

PLAIN MILLING MACHINE.

Patented Oct. 18, 1892; Jan. 18, 1898;
Feb. 6, 1900.

Fig.29

No. 12

26 in. x 5-8 in. x 7 1-2 in.

PLAIN MILLING MACHINE.

Fig.30

No. 13

15 in. x 3 in. x 9 3-4 in.

PLAIN MILLING MACHINES.

Fig.31

No. 24

72 in. x 12 in. x 19 in.

PLAIN MILLING MACHINE.

Patented February 6, 1894.

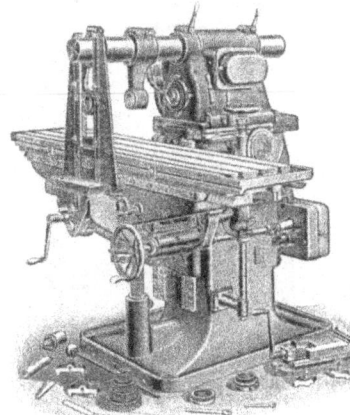

Fig.32

No. 2

26 in. x 12 in. x 20 in.

VERTICAL SPINDLE MILLING
MACHINE.

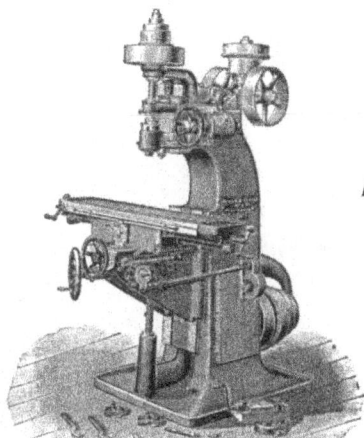

Fig.33

No. 5

52 in. x 12 in. x 21 1-2 in.

VERTICAL SPINDLE MILLING
MACHINE.

Patented February 6, 1900.
(Prior to 1900, No. 2.)

Fig.34

HIGH SPEED MILLING ATTACHMENTS.

For Nos. 1, 1 1-2, 2, 3 and 4 Universal, and Nos. 1, 2, 3 and 4 Plain Milling Machines.

Patented Feb. 6, 1900.

Fig.35

VERTICAL SPINDLE MILLING ATTACHMENTS

For Nos. 1, 1 1-2, 2 and 3 Universal, and Nos. 0, 1, 2 and 3 Plain Milling Machines.

Fig.36

SLOTTING ATTACHMENTS.

For Nos. 1, 1 1-2, 2, 3 and 4 Universal and Nos. 0, 1, 2, 3 and 4 Plain Milling Machines.

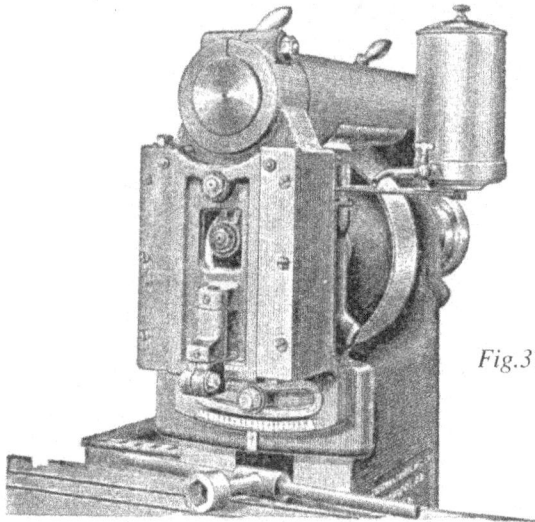

Fig.37

UNIVERSAL MILLING ATTACHMENT No. 12.

For Use on No. 4 Universal, and Nos. 4 and 24 Plain Milling Machines.

Fig.38

RACK CUTTING ATTACHMENTS.

Fig.39

10 Inch UNIVERSAL INDEX CENTRES.

Patented Feb. 5, 1884; Feb. 14, 1893.

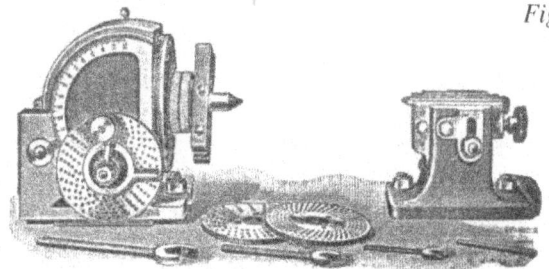

Fig.40

A number of milling machine improvements were patented in 1903, 1905, and 1906. These were reflected in the 1907 catalog listings that included universal milling machines in No.1 (Fig.41), No.2 (Fig.42), No.2A (Fig.43) No.3 (Fig.44), No.3A heavy (Fig.45), and No.4A heavy (Fig.46); plain milling machines in No.0 (Fig.47), No.0Y rack feed (Fig.48), No.1 (Fig.49), No.1Y rack feed (Fig.50), No.2 (Fig.51), No.2Y rack feed (Fig.52), No.2 heavy (Fig.53), No.3 (Fig.54), No.3B heavy (Fig.55), No.4B heavy (Fig.56), and No.5B heavy (Fig.57); vertical milling machines in No.2 (Fig.58), No.3 (Fig.59) and No.5 (Fig.60) sizes; vertical spindle milling attachments (Fig.61); compound vertical spindle milling attachments (Fig.62); cam cutting attachments (Fig.63); and gear cutting attachments (Fig.64).

No. 1

20 in. x 7 in. x 18 in.

UNIVERSAL MILLING MACHINE.

Patented Feb. 14, May 23, 1893; Aug. 29, 1899; Feb. 6, 1900;
Nov. 12, 1901; Jan. 13, 1903; July 11, Sept. 5,
Dec. 12, 1905.

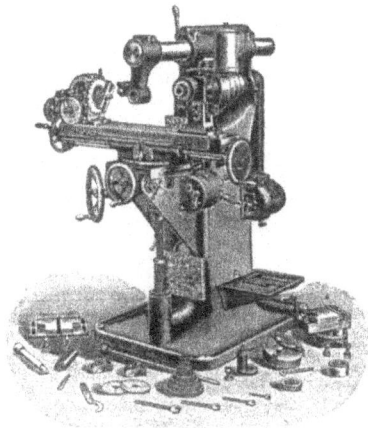

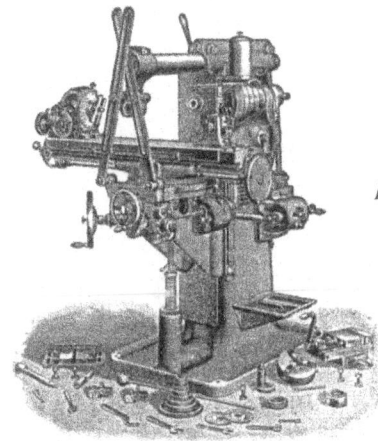

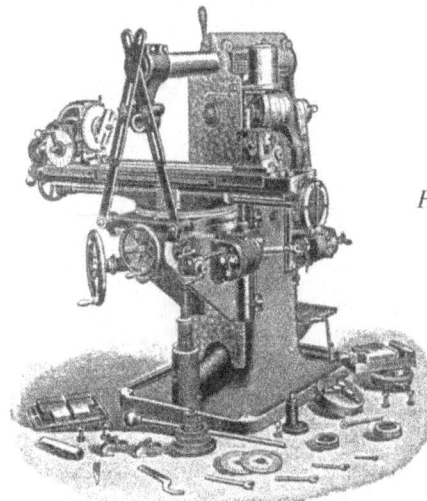

Fig.41

No. 2

25 in. x 8 in. x 18 in.

UNIVERSAL MILLING MACHINE.

Hand or Power Vertical Feed.

Patented Feb. 14, May 23, 1893; Aug. 29, 1899; Feb. 6, 1900;
Nov. 12, 1901; Jan. 13, 1903; July 11, Sept. 5,
Dec. 12, 1905.

Fig.42

No. 2-A

25 in. x 8 in. x 18 in.

UNIVERSAL MILLING MACHINE.

Constant Speed Drive.

Hand or Automatic Vertical Feed.

Patented Feb. 14, May 23, 1893; Aug. 29, 1899; Feb. 6, 1900;
Nov. 12, 1901; Jan. 13, 1903; July 11, Sept. 5, Dec. 12, 1905.
Others pending.

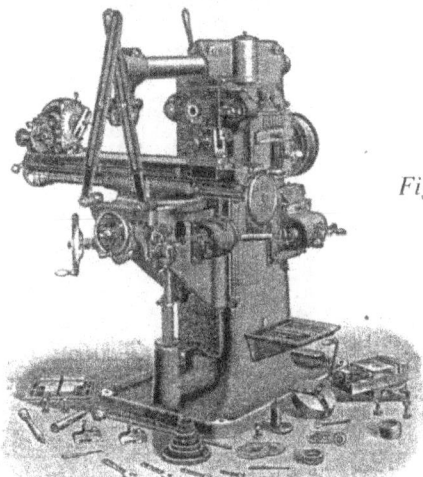

Fig.43

No. 3

30 in. x 10 in. x 19 in.

UNIVERSAL MILLING MACHINE.

Patented Feb. 14, May 23, 1893; Aug. 29, 1899; Feb. 6, 1900;
Nov. 12, 1901; Jan. 13, 1903; July 11, Sept. 5,
Dec. 12, 1905.

Fig.44

No. 3-A HEAVY
30 in. x 10 in. x 19 in.
UNIVERSAL MILLING MACHINE.
Constant Speed Drive.
Patented Feb. 14, May 23, 1893; Aug. 29, 1899; Feb. 6, 1900;
Nov. 12, 1901; Jan. 13, 1903; July 11, Sept. 5, Dec. 12, 1905.
Others pending.

Fig.45

18 in. x 6 in. x 15 in.

PLAIN MILLING MACHINE.

Patented May 23, 1893; Aug. 29, 1899, Jan. 13, 1903.

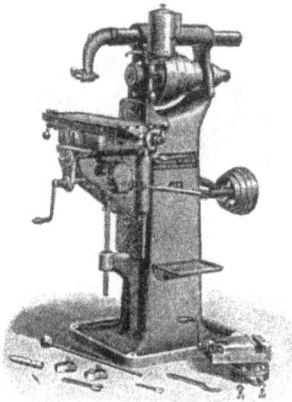

Fig.47

No. 1
24 in. x 7 in. x 19 in.
PLAIN MILLING MACHINE.
Patented May 23, 1893; Aug. 29, 1899; Feb. 6, 1900;
Jan. 13, 1903; July 11, Dec. 12, 1905.

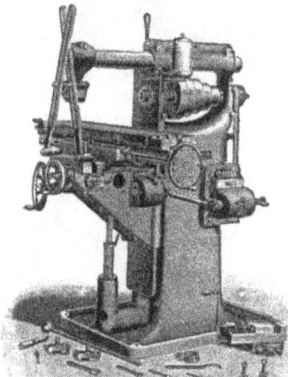

Fig.49

No. 4-A HEAVY
35 in. x 12 in. x 20 in.
UNIVERSAL MILLING MACHINE.
Constant Speed Drive.
Patented Feb. 14, May 23, 1893; Aug. 29, 1899; Feb. 6, 1900;
Nov. 12, 1901; Jan. 13, 1903; July 11, Sept. 5, Dec. 12, 1905.
Others pending.

Fig.46

No. 0-Y
18 in. x 6 in. x 15 in.
PLAIN MILLING MACHINE.
Rack Feed.

Fig.48

No. 1-Y
24 in. x 6 1-2 in. x 18 1-2 in.
PLAIN MILLING MACHINE.
Rack Feed.
Patented Dec. 12, 1905

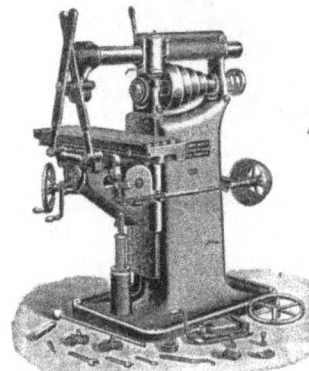

Fig.50

No. 2

28 in. x 8 in. x 19 in.

PLAIN MILLING MACHINE.

Hand or Automatic
Transverse and Vertical Feeds.

Patented May 23, 1893; Aug. 29, 1899; Jan. 13, 1903.
Others pending.

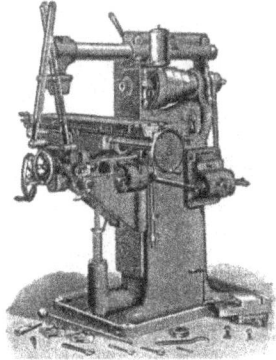

Fig.51

No. 2-Y

28 in. x 6 1-2 in. x 18 1-2 in.

PLAIN MILLING MACHINE.

Rack Feed.

Patented Dec. 12, 1905.

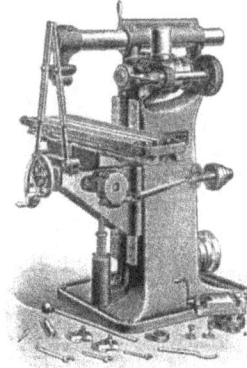

Fig.52

No. 2 HEAVY

28 in. x 8 in. x 19 in.

PLAIN MILLING MACHINE.

Hand or Automatic
Transverse and Vertical Feeds.

Patented May 23, 1893; Aug. 29, 1899; Jan. 13, 1903.
Others pending.

Fig.53

No. 3

34 in. x 10 in. x 20 in.

PLAIN MILLING MACHINE.

Hand or Automatic
Transverse and Vertical Feeds.

Patented Oct. 18, 1892; May 23, 1893; Aug. 29, 1899; Feb. 6, 1900;
Jan. 13, 1903; July 11, Dec. 12, 1905.

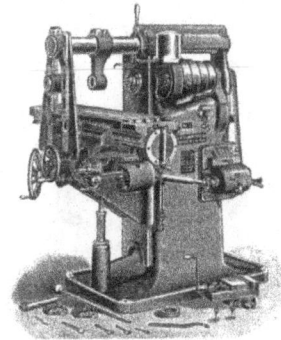

Fig.54

No. 3-B HEAVY

34 in. x 10 in. x 20 in.

PLAIN MILLING MACHINE.

Constant Speed Drive.
Hand or Automatic Transverse and Vertical Feeds.

Patented May 23, 1893; Feb. 6, 1900; Jan. 13, 1903; July 11,
Dec. 12, 1905.   Others pending.

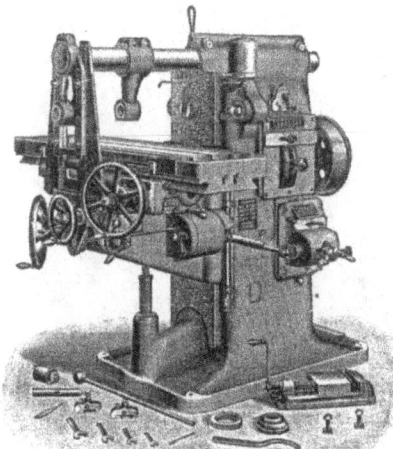

Fig.55

No. 4-B HEAVY

42 in. x 12 in. x 20 in.

PLAIN MILLING MACHINE.

Constant Speed Drive.
Hand or Automatic Transverse and Vertical Feeds.

Patented Oct. 18, 1892; May 23, 1893; Feb. 6, 1900;
Jan. 13, 1903; July 11, Dec. 12, 1905.
Others pending.

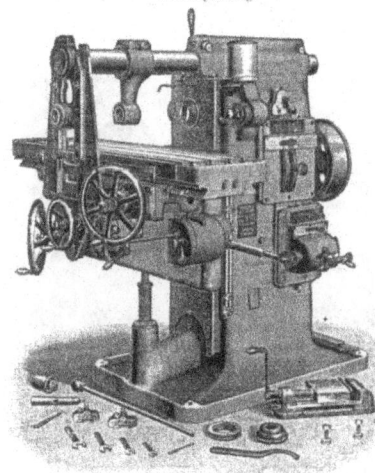

Fig.56

No. 5-B HEAVY
50 in. x 12 in. x 21 in.
PLAIN MILLING MACHINE.
Constant Speed Drive.
Patented Oct. 18, 1892; May 23, 1893; Feb. 6, 1900;
Jan. 13, 1903; July 11, 1905.

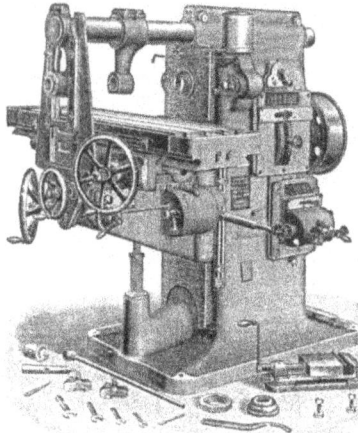

Fig.57

No. 2
28 in. x 12 in. x 20 in.
VERTICAL SPINDLE MILLING
MACHINE.
Patented May 23, 1893; Feb. 6, 1900; Jan. 13, 1903;
July 11, 1905; Jan. 9, 1906.

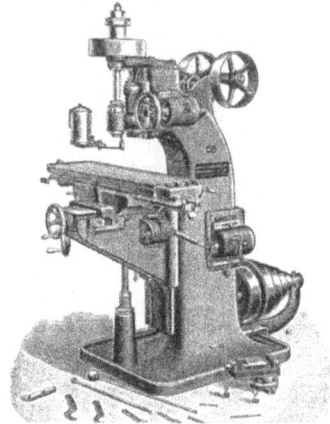

Fig.58

No. 3
34 in. x 13 1-2 in. x 23 in.
VERTICAL SPINDLE MILLING
MACHINE.
Constant Speed Drive.
Patented May 23, 1893; Feb. 6, 1900; Jan. 13, 1903;
July 11, 1905.   Others pending.

Fig.59

No. 5
52 in. x 12 in. x 24 in.
VERTICAL SPINDLE MILLING
MACHINE.
Patented Feb. 6, 1900; July 11, 1905.

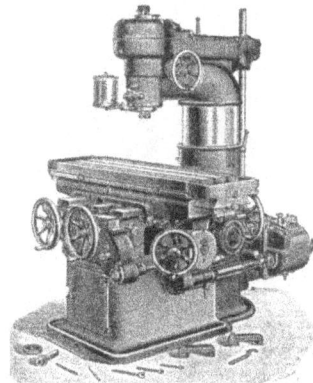

Fig.60

Nos. 20 and 2
COMPOUND VERTICAL SPINDLE
MILLING   ATTACHMENTS.
Patented July 10, 1906.

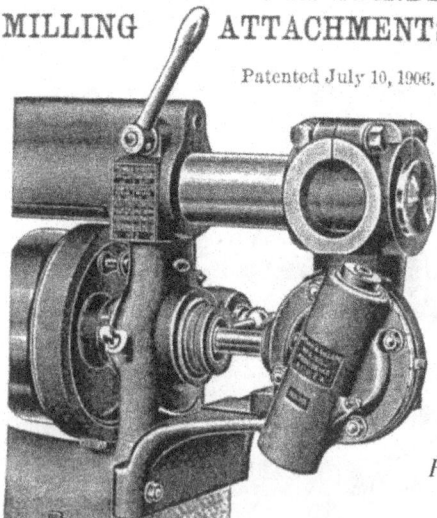

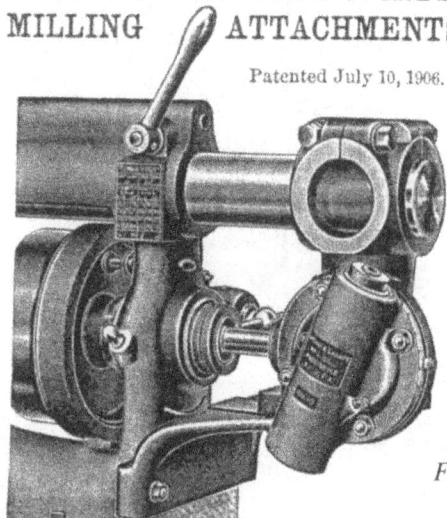

Fig.61

Nos. 20 and 2
COMPOUND VERTICAL SPINDLE
MILLING   ATTACHMENTS.
Patented July 10, 1906.

Fig.62

No. 10
CAM CUTTING ATTACHMENT.
For use on Nos. 1, 1 1-2, 2, 2A and 3 Universal,
and Nos. 1, 1Y, 1 1-2, 2, 2Y, 2 Heavy
and 3 Plain Milling Machines.

Fig.63

GEAR CUTTING ATTACHMENT.
Patented Sept. 5, 1905.

Fig.64

Further milling machine improvements were patented in 1912, 1916, and 1918. These were reflected in the 1920 catalog that included universal milling machines in No.1A (Fig.65), No.2 (Fig66), No.2A (Fig.67), No.3 (Fig.68), No 3A heavy (Fig.69), and No.4A heavy (Fig.70) sizes; hand milling machines in No.00 (Fig.71) size; plain milling machines in No.0 (Fig.72), No.1B (Fig.73), No.2B (Fig.74), No.2B heavy (Fig.75), No.3 (Fig.76), No.3B (Fig.77), No.3B heavy (Fig.78), No.4B (Fig.79), No.4B heavy (Fig.80), No.5B heavy (Fig.81), No.13B (Fig.82), and No.21 (Fig.83) sizes; vertical milling machines in No.2 (Fig.84) and No.3 (Fig.85) sizes; vertical spindle milling attachments (Figs.86-87); universal milling attachments (Fig.88); slotting attachments (Fig.89); rack cutting attachments (Figs.90-91); cam cutting attachments (Fig.92); spiral milling attachments (Fig.93); short lead attachments (Fig.94); and horizontal milling attachments (Fig.95).

No. 1A
22 in. x 8 in. x 18 in.
UNIVERSAL MILLING MACHINE
(With Back Gears)
Constant Speed Drive

Patented Jan. 6, Jan. 13, 1903; July 11, Sept. 5, Dec. 12, 1905;
April 30, May 28, 1907; July 16, 1912; Feb. 1,
Sept. 5, 1916. Patent pending.

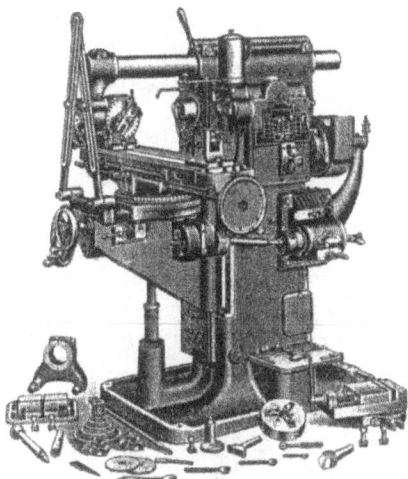

Fig.65

No. 2
28 in. x 10 in. x 18 in.
UNIVERSAL MILLING MACHINE

Patented Jan. 13, 1903; July 11, Sept. 5, Dec. 12, 1905;
April 30, 1907; Dec. 3, 1918.

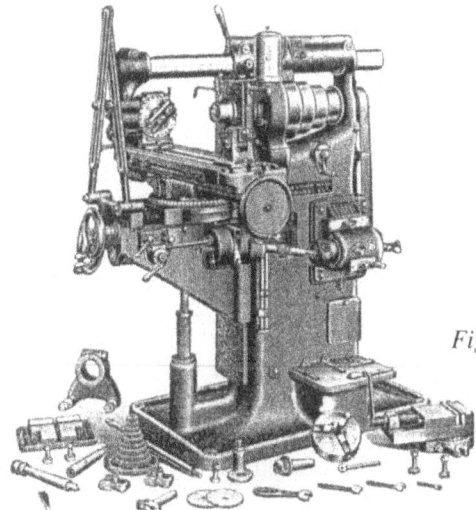

Fig.66

54

No. 2A

28 in. x 10 in. x 18 in.

UNIVERSAL MILLING MACHINE

Constant Speed Drive

Patented Jan. 6, Jan. 13, 1903; July 11, Sept. 5, Dec. 12, 1905;
April 30, May 28, 1907; July 16, 1912; Feb. 1,
Sept. 5, 1916; Dec. 3, 1918.

Fig.67

No. 3

34 in. x 12 in. x 19 in.

UNIVERSAL MILLING MACHINE

Patented Jan. 13, 1903; July 11, Sept. 5, Dec. 12, 1905;
April 30, 1907; Dec. 3, 1918;
Feb. 25, 1919.

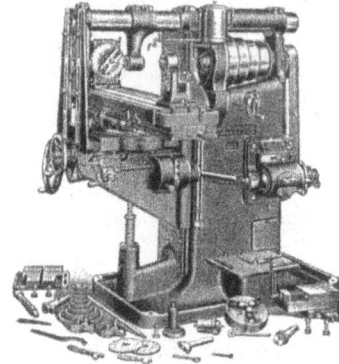

Fig.68

No. 3A HEAVY

34 in. x 12 in. x 19 in.

UNIVERSAL MILLING MACHINE

Constant Speed Drive

Patented Jan. 6, Jan. 13, 1903; July 11, Sept. 5, Dec. 12, 1905;
April 30, May 28, 1907; July 16, 1912; Feb. 1, Sept. 5,
1916; Dec. 3, 1918; Feb. 25, 1919.

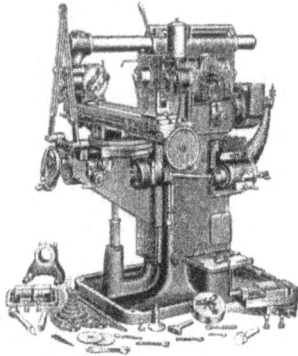

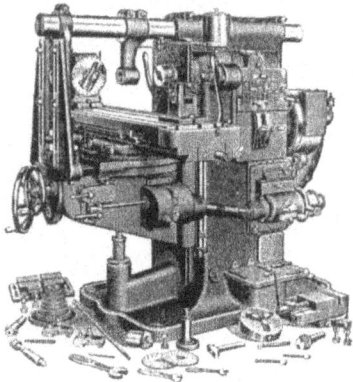

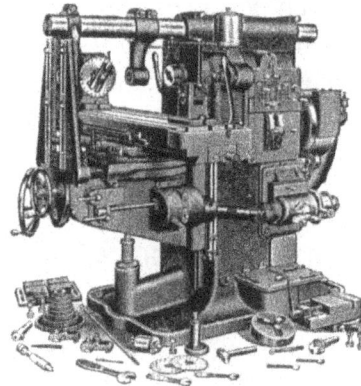

Fig.69

No. 4A HEAVY

42 in. x 14 in. x 20 in.

UNIVERSAL MILLING MACHINE

Constant Speed Drive

Patented Jan. 6, Jan. 13, 1903; July 11, Sept. 5, Dec. 12, 1905;
April 30, May 28, 1907; July 16, 1912; Feb. 1, Sept. 5,
1916; Dec. 3, 1918; Feb. 25, 1919.

Fig.70

No. 00

7 in. x 4 1-4 in. x 7 1-2 in.

HAND MILLING MACHINE

All Hand Feeds

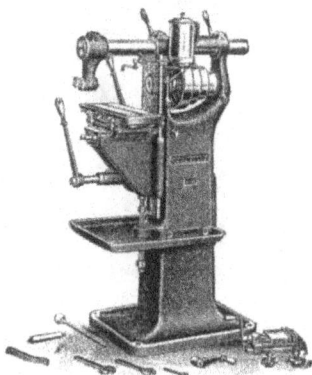

Fig.71

No. 0

18 in. x 6 in. x 15 in.

PLAIN MILLING MACHINE

Screw Feed

Patented Jan. 13, 1903.

Fig.72

No. 1B

22 in. x 8 in. x 19 in.

PLAIN MILLING MACHINE

(With Back Gears)

Constant Speed Drive

Patented Jan. 6, Jan. 13, 1903; July 11, Dec. 12, 1905;
May 28, 1907; July 16, 1912; Feb. 1,
Sept. 5, 1916.

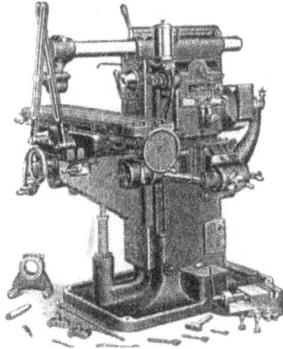

Fig.73

No. 2B

28 in. x 10 in. x 19 in.

PLAIN MILLING MACHINE

Constant Speed Drive

Patented Jan. 6, Jan. 13, 1903; July 11, Dec. 12, 1905;
May 28, 1907; July 16, 1912; Feb. 1, Sept. 5, 1916;
Dec. 3, 1918.

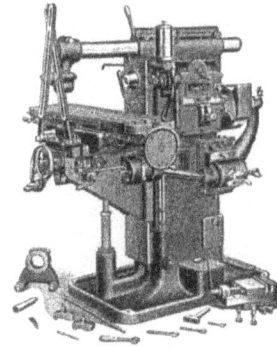

Fig.74

No. 2B HEAVY

28 in. x 10 in. x 19 in.

PLAIN MILLING MACHINE

Constant Speed Drive

Patented Feb. 6, 1900; Jan. 6, Jan. 13, 1903; July 11, Dec. 12,
1905; May 28, 1907; July 16, 1912; Feb. 1,
Sept. 5, 1916; Feb. 25, 1919.

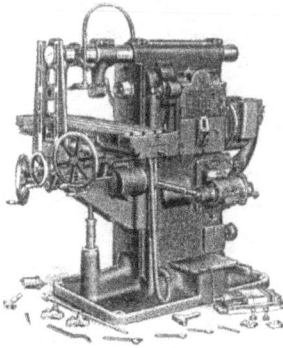

Fig.75

No. 3

34 in. x 12 in. x 20 in.

PLAIN MILLING MACHINE

Patented Jan. 13, 1903; July 11, 1905; Dec. 3, 1918;
Feb. 25, 1919.

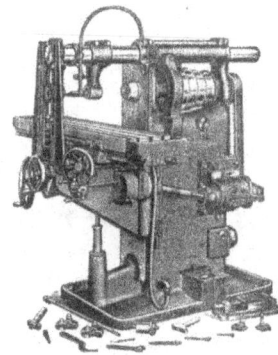

Fig.76

No. 3B

34 in. x 12 in. x 20 in.

PLAIN MILLING MACHINE

Constant Speed Drive

Patented Feb. 6, 1900; Jan. 6, Jan. 13, 1903; July 11, Dec. 12,
1905; May 28, 1907; July 16, 1912; Feb. 1,
Sept. 5, 1916; Feb. 25, 1919.

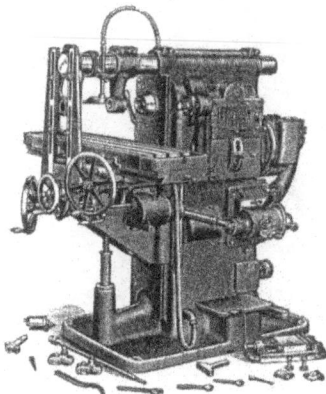

Fig.77

No. 3B HEAVY

34 in. x 12 in. x 20 in.

PLAIN MILLING MACHINE

Constant Speed Drive

Patented Jan. 6, Jan. 13, 1903; July 11, Dec. 12, 1905;
May 28, 1907; July 16, 1912; Feb. 1, Sept. 5, 1916;
Dec. 3, 1918; Feb. 25, 1919.

Fig.78

No. 4B
42 in. x 14 in. x 20 in.
PLAIN MILLING MACHINE
Constant Speed Drive
Patented Jan. 6, Jan. 13, 1903; July 11, Dec. 12, 1905;
May 28, 1907; July 16, 1912; Feb. 1, Sept. 5, 1916;
Dec. 3, 1918; Feb. 25, 1919.

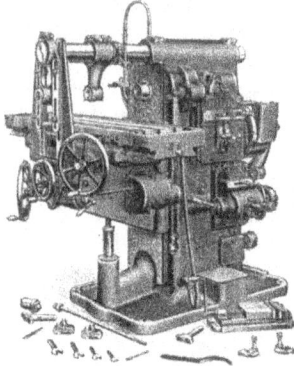

Fig. 79

No. 4B HEAVY
42 in. x 14 in. x 20 in.
PLAIN MILLING MACHINE
Constant Speed Drive
Patented Jan. 6, Jan. 13, 1903; July 11, Dec. 12, 1905;
May 28, 1907; July 16, 1912; Feb. 1, Sept. 5, 1916;
Dec. 3, 1918; Feb. 25, 1919.

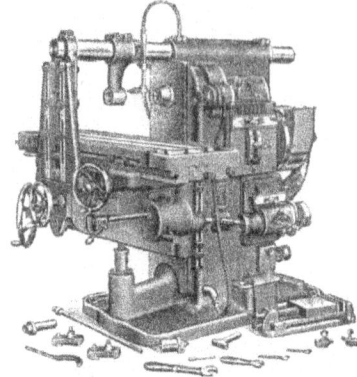

Fig. 80

No. 5B HEAVY
50 in. x 14 in. x 21 in.
PLAIN MILLING MACHINE
Constant Speed Drive
Patented Jan. 6, Jan. 13, 1903; July 11, 1905; May 28, 1907;
July 16, 1912; Feb. 1, Sept. 5, 1916; Dec. 3, 1918;
Feb. 25, 1919.

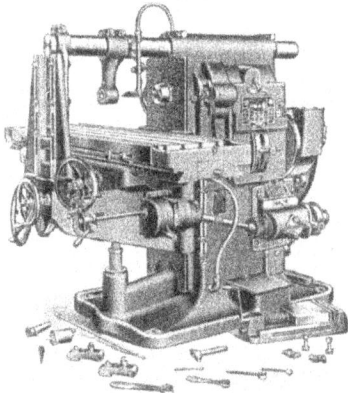

Fig. 81

No. 13B
34 in. x 5 1-2 in. x 12 in.
PLAIN MILLING MACHINE.
Constant Speed Drive.

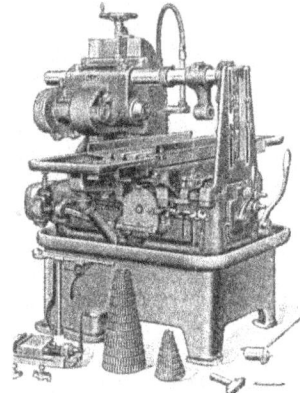

Fig. 82

No. 21
22 in. x 5 3-4 in. x 14 1-2 in.
AUTOMATIC MILLING MACHINE
Constant Speed Drive
Patented Jan. 6, 1903; Dec. 12, 1905. Patent pending.

Fig. 83

No. 2
28 in. x 14 in. x 20 in.
VERTICAL SPINDLE MILLING
MACHINE
Constant Speed Drive
Patented Jan. 6, Jan. 13, 1903; July 11, 1905; May 28, 1907;
Sept. 5, 1916; Feb. 25, 1919.

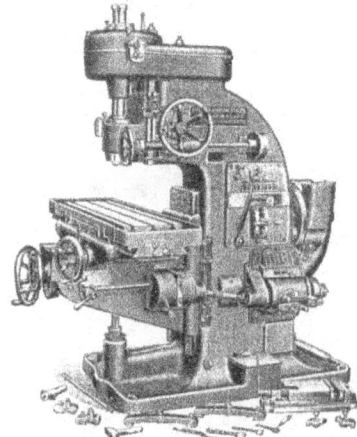

Fig. 84

No. 3
34 in. x 16 in. x 22 in.

VERTICAL SPINDLE MILLING MACHINE
Constant Speed Drive

Patented Jan. 6, Jan. 13, 1903; July 11, 1905; May 28, 1907; Sept. 5, 1916; Dec. 3, 1918; Feb. 25, 1919.

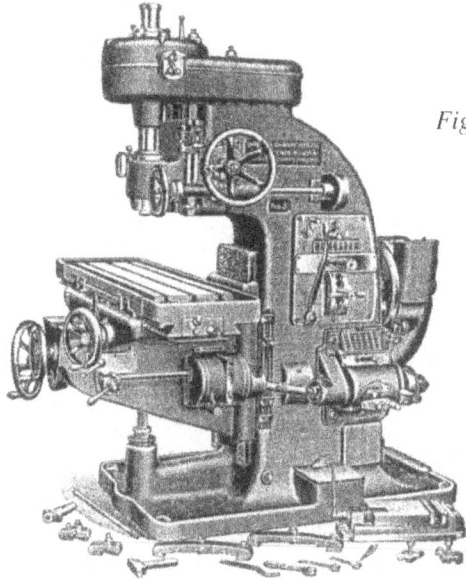

Fig.85

Nos. 0, 1 and 2
VERTICAL SPINDLE MILLING ATTACHMENTS

When ordering, give size and serial number of machine

Fig.86

Nos. 1H, 2H, 3, 4 & 5
VERTICAL SPINDLE MILLING ATTACHMENTS

Fig.87

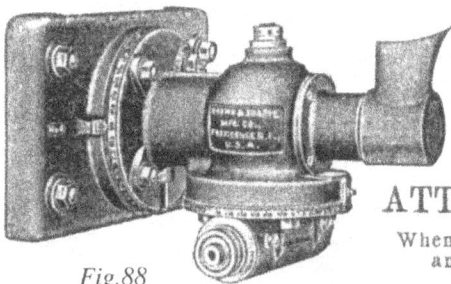

Nos. 1, 2, 3, 4
UNIVERSAL MILLING ATTACHMENTS

When ordering, give size and serial number of machine

Fig.88

Nos. 0, 1, 2, 3, 4 & 5
SLOTTING ATTACHMENTS
When ordering, give size and serial number of machine

Fig.89

Nos. 1, 2, 3 and 4
RACK CUTTING ATTACHMENTS

When ordering, give size and serial number of machine

Fig.90

INDEXING ATTACHMENTS
ENGLISH OR METRIC
For Use with Rack Cutting Attachments

When ordering, give size and serial number of machine; also whether English or Metric Attachment is desired

Fig.91

No. 10
CAM CUTTING ATTACHMENT

For use on Nos. 1, 1A (with or without Back Gears), 2, 2A,
2A Heavy, 3 and 3A Universal, and Nos. 1, 1B (with or
without Back Gears), 1Y, 2, 2Y, 2B, 2B Heavy,
3 and 3B Plain Milling Machines

Fig.92

Nos. 1, 2, 3 & 4
SPIRAL MILLING ATTACHMENTS

When ordering, give size
and serial number of
machine

Fig.93

No. 2 SHORT LEAD ATTACHMENT

For Nos. 1, 1A (With or without Back Gears),
2 and 2A Universal Milling Machines

Fig.94

No. 1 HORIZONTAL MILLING ATTACHMENT

*For No. 1 Vertical Spindle Milling Mch.

Fig.95

BUCKEYE IRON & BRASS WORKS, Dayton, OH

Organized by Charles E. Pease (1836-1910) in 1876 to make machine tools and tobacco machinery. Milling machine production began with a design patented August 17, 1875 by William Krutzsch and *shown at right* as offered in 1880. It featured back gears on the spindle, a rather complex lead screw feed to the table, saddle and knee motions, and a coolant sump in the base for supplying coolant to the cutters by means of a rotary pump. A gear cutting attachment, also patented August 17, 1875, by William Krutzsch was offered with the machine. A multiple milling machine, patented May 15, 1883, by Krutzsch, may have also been built.

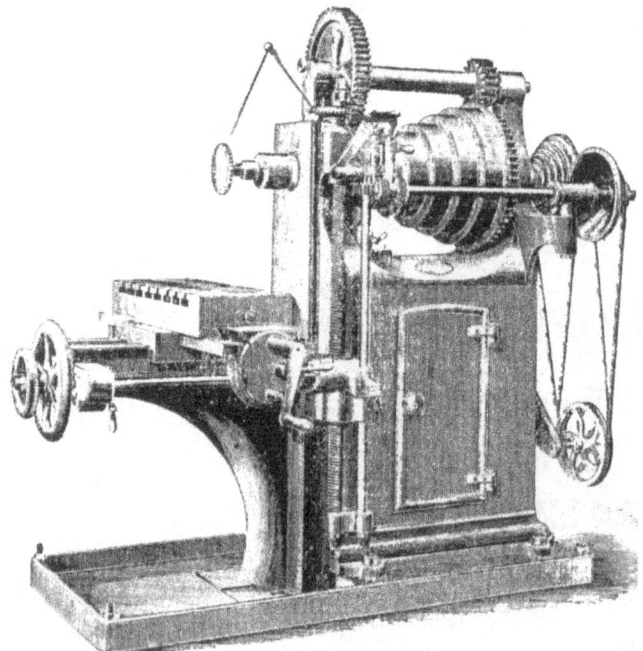

BURKE MACHINERY CO., Cleveland, OH,
later BURKE MACHINE TOOL CO., Conneaut, OH

Founded about 1905 by George H. Burke (1878-1933) to make small machine tools. In 1910, Burke moved the company to Conneaut, OH, and reorganized as the Burke Machine Tool Co. Milling machine production included two-spindle bench millers (Fig.1) introduced in 1907, with table travel of 6" and 2½" cross travel. No.4 milling machines (Fig.2) were also advertised in 1907, with a smaller No.3 hand feed machine (Fig.3) appearing by 1916. A vertical milling machine (Fig.4) with hand feed and a 15" x 5¾" table was introduced in 1918. The firm continued production of small machines until 1948 when it was taken over by the Smith & Mills Co.

When The Work Comes In Bunches

That's the time our New Two Spindle Milling Machine proves its worth. It's a fast worker and handles a large variety of work with ease and marked economy. Mills two ends of a piece at the same time and insures true and accurate results. Get a circular which describes this leader in detail.

THE BURKE MACHINERY CO., Cleveland, O.

Fig.1

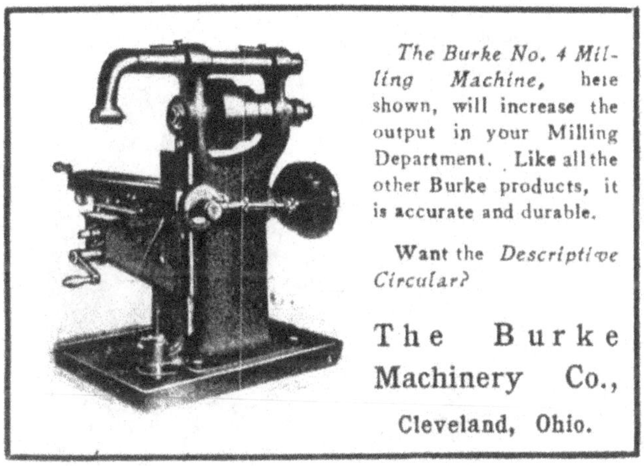

The Burke No. 4 Milling Machine, here shown, will increase the output in your Milling Department. Like all the other Burke products, it is accurate and durable.

Want the *Descriptive Circular?*

The Burke Machinery Co., Cleveland, Ohio.

Fig.2

Burke's No. 3
Bench Millers
Are Seasoned Tools

Our system of tests insures accuracy, and our methods of construction make it permanent. Frames and parts are carried 6 to 8 months, after work begins on them, before the completed tools receive our final O. K., and all parts are finished and fitted with exceptional care.

Longitudinal feed, 6 in.; traverse feed, 3¼ in.; vertical motion of knee, 8¼ in.; table surface, 3½x12 inch.

Write for Catalog, which also shows larger and smaller sizes.

The Burke Machine Tool Co.
Conneaut, Ohio

Fig.3

BURKE
Vertical Bench Millers

will handle a great many of these small vertical milling operations which you have in your factory. Let us show you. We also build plain bench milling machines with hand or power feed. Want our catalog?

Fig.4

BURR & SONS, JOHN T., Brooklyn, NY

Founded by John T. Burr (1840-1916) about 1890 to make keyseating machines which he had patented in 1885. Burr operated the company until his death in 1916. Keyseat milling machines, which could cut keyseats in shafts from ¾" to 4" diameter and 20" long, were offered beginning in 1896 (Fig.1) A larger version (Fig.2) was introduced in 1901 and a combination keyseat milling and routing machine (Fig.3) in 1903. The combination machine was equipped with a vertical spindle for cutting keyways in shaft ends.

KEY-SEAT MILLING MACHINE.

Fig.1

KEYSEAT MILLING MACHINE

Fig.2

Fig.3

BURR KEY SEAT MILLING AND ROUTING MACHINE.

CAMPBELL, A.H., Hartford, CT

Maker, in 1881, of a portable, hand powered, milling machine for milling locomotive steam chests in railroad repair shops, *see right*.

By 1884, the device was a product of PEDRICK & AYER.

STEAM CHEST SEAT MILLING MACHINE.

CARTER & HAKES MACHINE CO.

Winsted, CT

Founded by Franklin P. Carter (1852-1923) and Curtis Hakes in 1899. Milling machines were the primary product. Production included No.3 plain milling machines (Fig.1) with 24" table travel and 5¾" cross travel and No.4 plain milling machines (Fig.2) with 30" table travel and 6¾" cross travel, both introduced in 1899; Lincoln type millers with quick return (Fig.3); small bench milling machines (Fig.4) with 4" table travel, 2¼" cross travel and 4" vertical travel, introduced by 1904 and offered into the 1920s; and No.3 hand millers (Fig.5) introduced in 1910. *(illustration continued next page)*

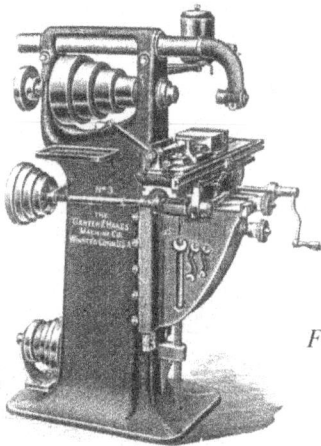

Fig.1

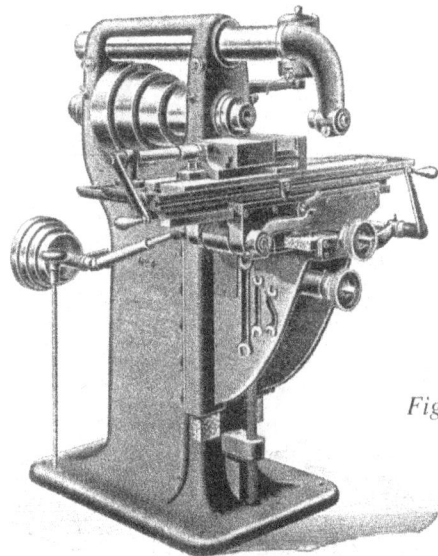

Fig.2

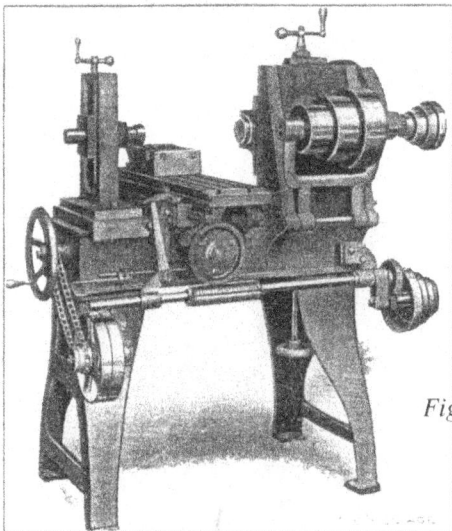

Fig.3

THE CARTER & HAKES QUICK RETURN MILLING MACHINE PLATEN

Bench Milling Machine.

Fig.4

Carter & Hakes Fig. 5

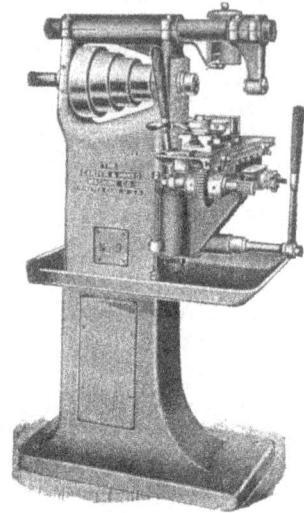

CHICAGO MACHINE TOOL CO., Chicago, IL

Formed in 1902, the company appears to have been controlled by Hill, Clarke & Co., a large Boston machinery dealer which acted as sole agent. Albert W. Wigglesworth, treasurer of Hill, Clarke, served as president until his death in 1917. The firm ceased operation in 1919. 1903 production was limited to hand milling machines equipped with both vertical and horizontal spindles. These included the Nos. 1 and 2 (Figs.1-2) bench type; No.2 floor type (Fig.3) with vertical spindle replaced by overarm; No.3 floor type with vertical hand horizontal spindles (Fig.4); and No.3 floor type with vertical spindle replaced by overarm (Fig.5). In 1907, the company introduced a bench miller with no provision for a vertical spindle (Fig.6). An improved column and means of vertical adjustment (Fig.7) was introduced in 1910.

(continued next page)

Nos. 1 and 2 Chicago Hand Millers

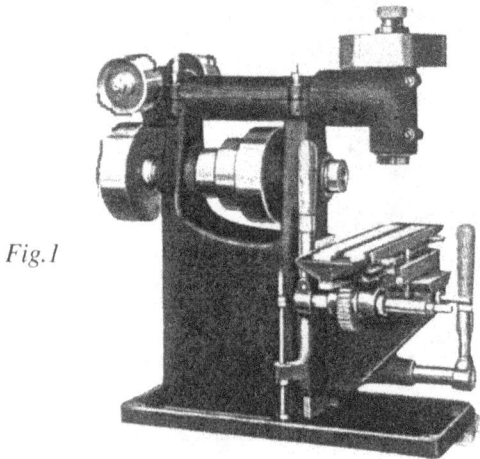

Fig.1

No. 2 Chicago Hand Miller
Vertical Spindle Replaced by Overhanging Arm

Fig.3

SIZES AND DIMENSIONS

SIZE	No. 1	No. 2	No. 3
Crank feed to table...........................	9"	10½"	15"
Lever feed to table..........................	4½"	5¾"	7"
Transverse feed.............................	2½"	3"	4"
Vertical motion of knee......................	4¼"	5¾"	7¼"
Greatest distance, top of table to center of spindle	4"	5"	7"
Working surface of table	3¼" x 10½"	4" x 12½"	5" x 20"
Largest cone diameter......................	5"	6"	8½"
Width of driving belt.......................	1½"	1¾"	2"
Taper hole in spindle.......................	B. & S. No. 9	B. & S. No. 9	B. & S. No. 9
Hole through spindle........................	¾"	¾"	¾"
Height over all.............................	22¾"	26¾"	53"
Weight, complete...........................	300 lbs.	440 lbs.	730 lbs.

Fig.2

No. 3 Chicago Hand Miller

No. 3 Chicago Hand Miller

Fig.4

Fig.5

Our New Bench Miller

The latest addition to our
family of Hand Milling
Machines

Two speed Self oiling countershaft
Drawing in collet
"Everlasting" self oiling journals
Capacity 9" x 2½"

Chicago
Machine Tool Co.,
CHICAGO   ILLINOIS

DOMESTIC AGENTS
HILL, CLARKE & CO., Inc.,
New York   Philadelphia
Boston   Chicago

Fig.6

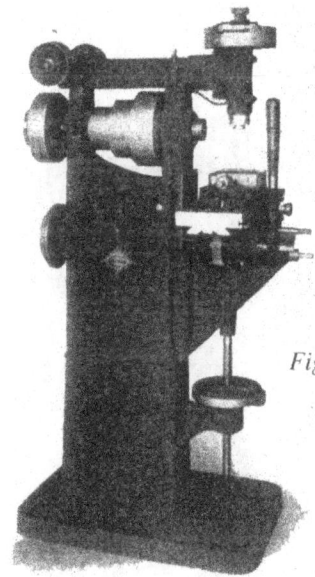

Fig.7

CINCINNATI MILLING MACHINE CO., Cincinnati, OH

Formed January 10, 1889, as a spin off of the milling machine business of the CINCINNATI SCREW & TAP CO. Frederick Holz (1853-1919) served a president until selling out to Frederick A. Geier (1866-1934) in 1905. Geier served as president from 1905 until his death in 1934.

The first model produced by the new company was a universal milling machine (Fig.1) that had been designed and prototyped at Cincinnati Screw & Tap Co. It was offered with the universal dividing head previously made by Cincinnati Screw & Tap. Late in 1889, an improved model of the universal milling machine (Fig.2) was introduced, accompanied by an improved universal dividing head (Fig.3). A new No.3 universal (Fig.4), with improvements patented July 22, 1890, was brought out by early 1891. The machine was equipped with 22" table travel, 7" cross travel and 18" vertical travel. 1892 production included a new No.5 plain miller (Fig.5) with 29" table travel; and an improved No.1 universal (Figs.6-7) with 18" table travel, 6" cross travel and 17" vertical travel.

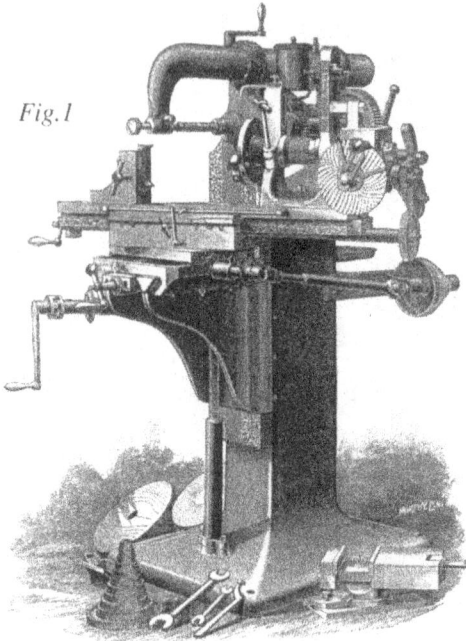

Fig.1

UNIVERSAL MILLING MACHINE.

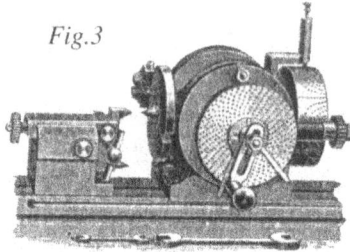

Fig.3

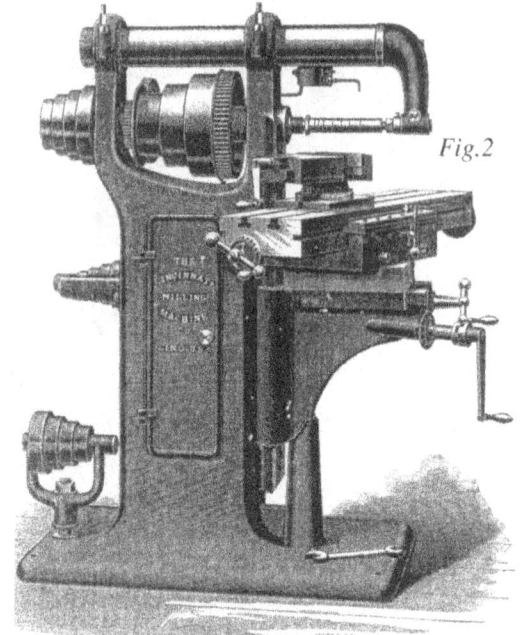

Fig.2

UNIVERSAL MILLING MACHINE.

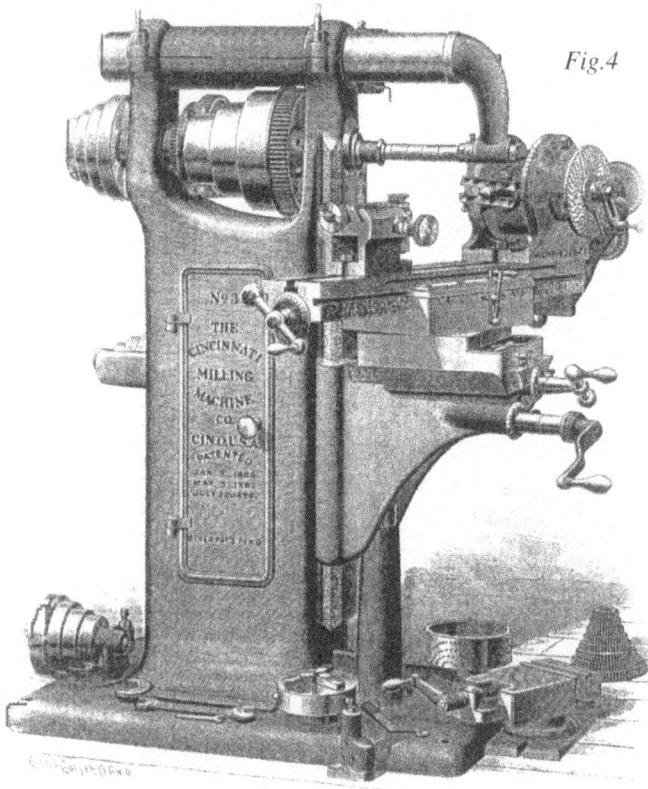

Fig.4

UNIVERSAL MILLING MACHINE.

Fig.5

NEW PLAIN MILLING MACHINE.

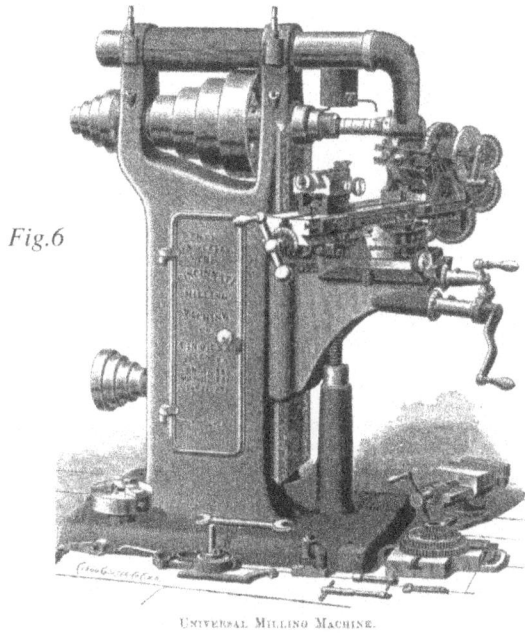

Fig.6

UNIVERSAL MILLING MACHINE.

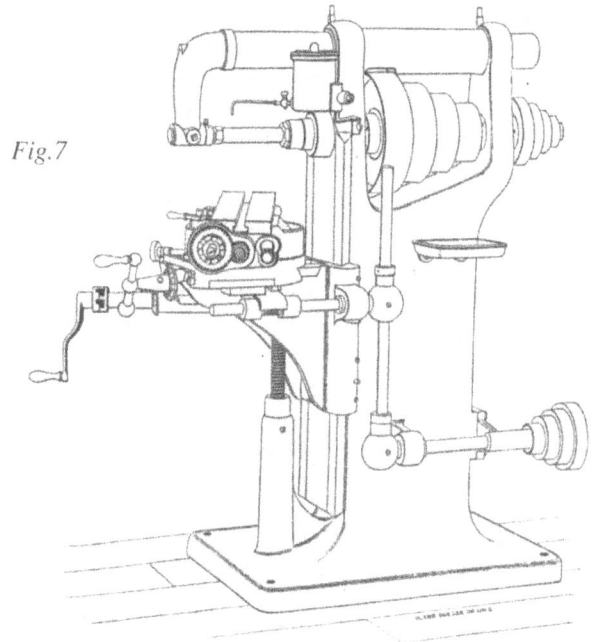

Fig.7

An improved line, with features patented January 30, 1894, was introduced in 1895. Models included a No.0 plain (Fig.8) with 18" table travel, 5½" cross travel and 14 vertical travel; No.2 plain (Fig.9) with 24" table travel, 7½" cross travel and 20" vertical travel; No.2 universal (Fig.10) with 23" table travel, 6½" cross travel and 18½" vertical travel; and No.3 universal (Fig.11) with 28" table travel, 7½" cross travel and 19" vertical travel.

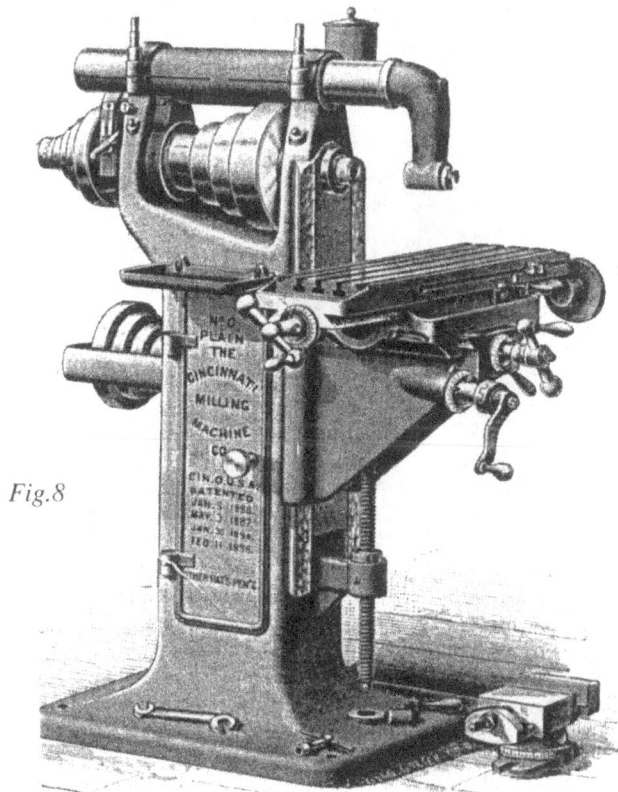

Fig.8

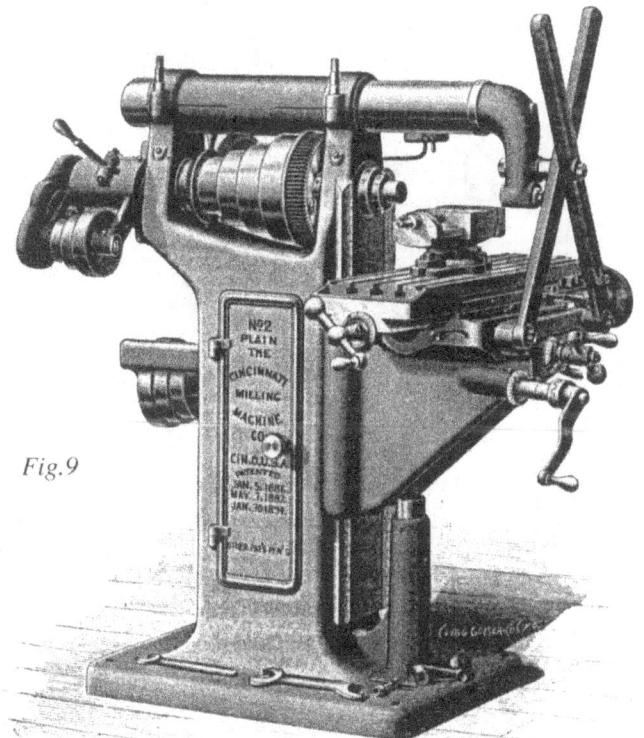

Fig.9

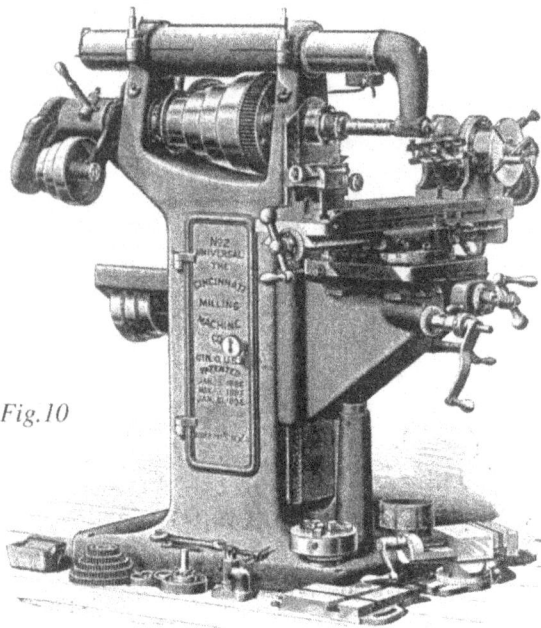

Fig.10

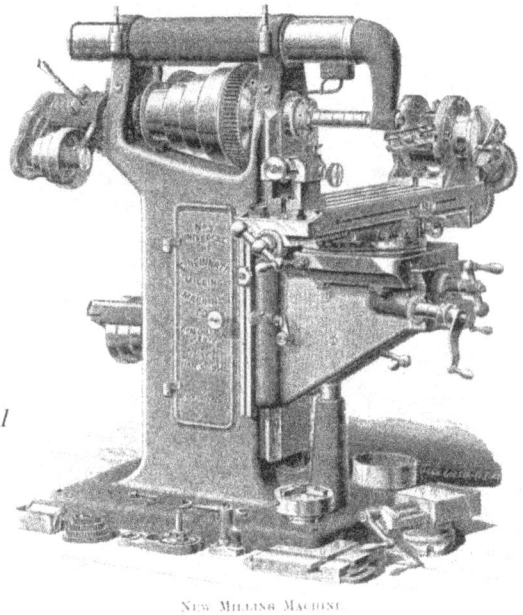

Fig.11

NEW MILLING MACHINE

A further improved line, patented October 2, 1900, with positive all-gear drive and improved overarm and brace design was introduced in 1900. A new No.4 plain miller (Fig.12) was the first to be brought out, followed by the No.3 universal (Fig.13) in 1901. New vertical milling attachments (Fig.14) were also offered in 1901 and new dividing heads (Fig.15) in 1903. By 1904, the machines were available with motor drive (Fig.16).

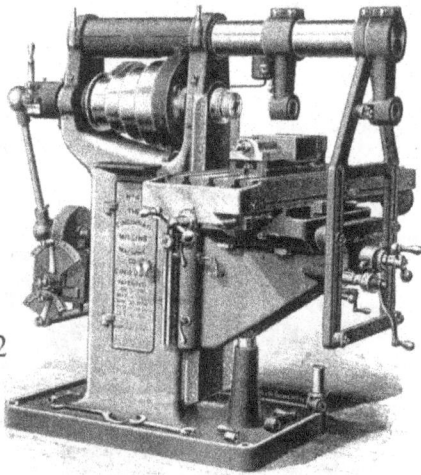

Fig.12

Fig.13

THE NEW CINCINNATI MILLING MACHINE

Fig.14

VERTICAL SPINDLE MILLING HEAD FOR CINCINNATI MILLER

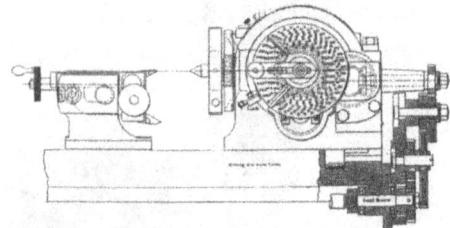

Fig.15

NEW INDEX HEAD FOR UNIVERSAL MILLING MACHINES

The line was completely redesigned in 1908, resulting in the HIGH POWER model. It was offered in horizontal spindle, constant speed models including No.2 plain (Fig.17) and No.2 universal (Fig.18); cone drive models including No.2 plain (Fig.19); motor drive models (Fig.20); and vertical spindle models with constant speed drive and rotary tables (Fig.21), offered in No.2, No.3 and No.4 sizes, with 28", 34" and 42" table travel respectively. A line of cone drive machines with the same geared feed but without spindle gear boxes, was also available by 1911. Models included No.2 plain (Fig.22) and No.2 universal (Fig.23). The last new HIGH POWER model, the No.5 plain (Fig.24), with 50" table travel, 14" cross travel, 21" vertical travel and box section overarm, was introduced in 1919. All the above models were offered into the 1920s and were gradually replaced by the M type.

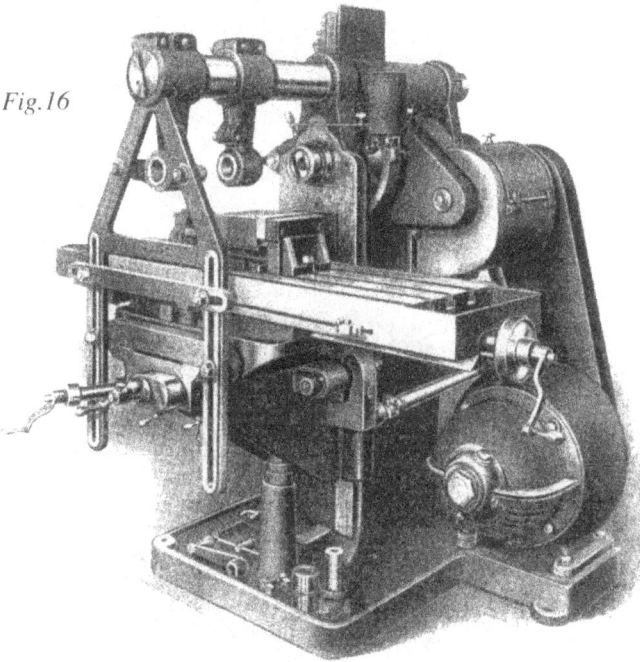

Fig.16

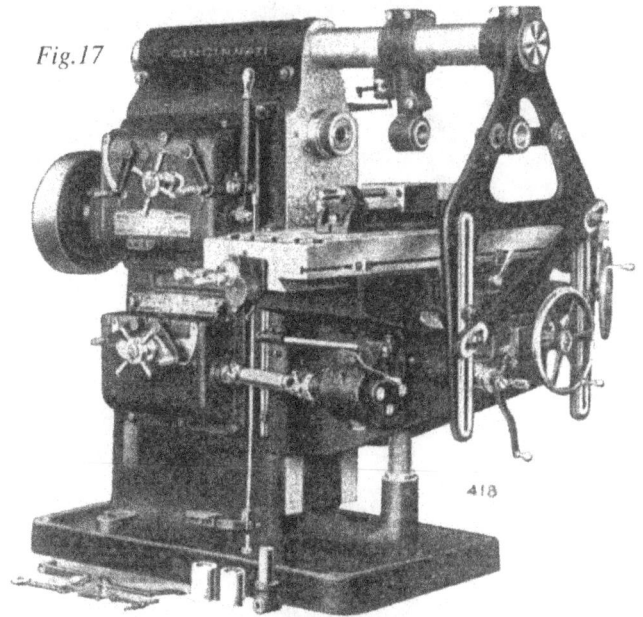

Fig.17

Fig.18

Fig.19

Fig 20

Fig.21

Fig.22

The No. 2 Plain

Fig.23

The No. 2 Universal

Fig.24

Fig.25

18" Cincinnati Automatic Milling Machine

Bed type millers, which owed much to the Lincoln design, were introduced in 1916. Early production was built in 12" plain (Fig.25), and 18" duplex; later production included 18", 24" and 48" sizes. 12" plain manufacturing millers (Fig.26) were introduced in 1917 to meet the great demand for simple machines dedicated to production of rifles and other small arms.

Fig.26

CINCINNATI 12-IN. PLAIN MANUFACTURING MILLER

CINCINNATI PUNCH & SHEAR CO.

Cincinnati, OH

Formed in 1901 as a reorganization of the Wais & Roos Punch & Shear Co. Production was generally limited to punching and shearing machinery. In 1909, the firm introduced a combination milling and boring machine, *shown at left*, with 48" of table travel and weighing 6,000 pounds. This machine appears to have been the only metal cutting machine tool built by Cincinnati Punch & Shear.

CINCINNATI SCREW & TAP CO.,

Cincinnati, OH

A partnership of Frederick Holz (1853-1919) and George A. Mueller (1854-1901) formed in 1874 as Holz & Mueller. The partners began operating as the Cincinnati Screw & Tap Co. in 1876 and incorporated in 1884. Holz served as president from 1887 to 1889. Milling machine production began in 1882 with the introduction of small milling machines (Fig.1) originally developed by Holz for making taps in their own shop. An improved version (Fig.2) with an overarm, was in production by 1883. A universal dividing head (Fig.3) patented January 5, 1886, by Holz; and milling machine centers (Fig.4) were introduced in 1886. The milling machine part of the business was spun off and incorporated as the CINCINNATI

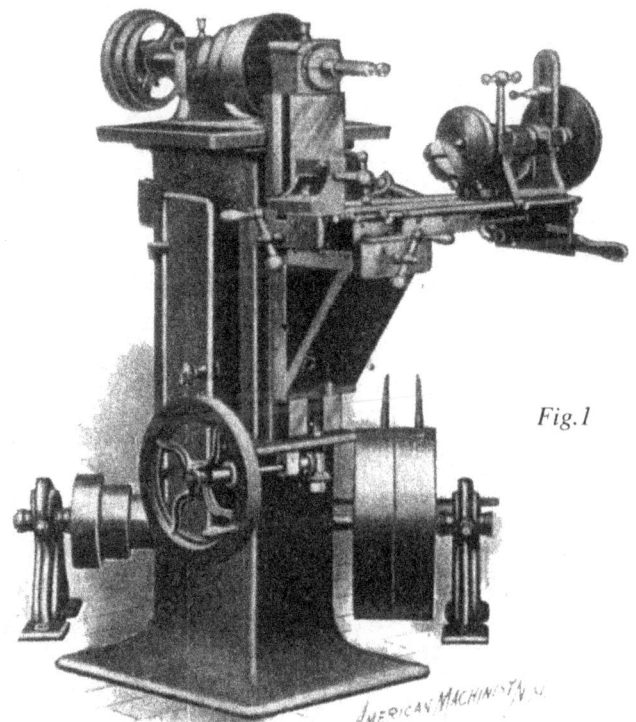

Fig.1

AMERICAN MACHINIST V

MILLING MACHINE CO. on January 10, 1889. The Cincinnati Screw & Tap Co. continued to operate until 1915.

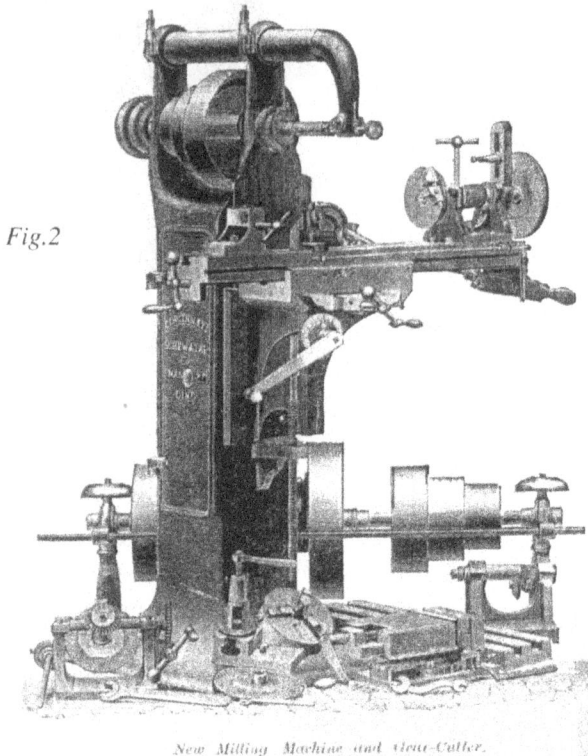

Fig.2

Fig.3

New Milling Machine and Gear-Cutter.

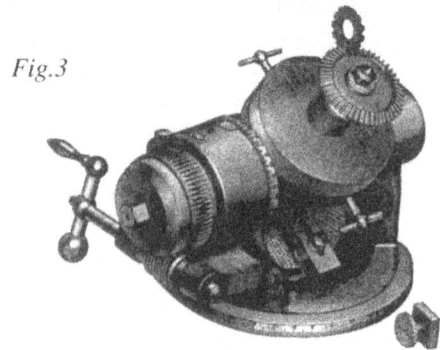

Fig.4

CENTERS FOR MILLING-MACHINES, PLANERS AND SHAPERS.

CLARK-MESKER CO., Cleveland, OH
See CLEVELAND MILLING MACHINE CO.

CLEVELAND MACHINE SCREW CO.,
Cleveland, OH

Formed in 1893 to make automatic screw machines, the firm was reorganized as the Cleveland Automatic Machine Co. in 1902. Its only non-screw machine product, the worm milling machine *shown at right*, was introduced in 1901.

New Worm Milling Machine.

Greatly reduces cost of making worms.
Skilled mechanic not necessary to run it.
One man can attend to five or six machines.

Mills worms of any pitch, single, double, triple or quadruple threads (either right or left hand.) Worms from one inch to four inches and all lengths up to six inches produced five to eight times faster than a lathe. Each worm absolutely accurate in pitch and angle.

Cuts even or uneven number of threads per inch without compounding change gear.

Write department "M" for full information and prices.

Cleveland Machine
        Screw Company,

Cleveland, Ohio, U. S. A.

CLEVELAND MILLING MACHINE CO., Cleveland, OH

Formed in 1916 by Frank S. Shields and John A. Camm (1880-1944). Camm, who left in 1918, had been with the KEARNEY & TRECKER CO. from 1909 to 1916. Its sole machine product appears to have been a No.2 milling machine, shown below, introduced in 1917 and made until the company failed in 1920. The same machine was made by the Clark-Mesker Co. from 1920 to 1923 and by the Shields Machine Tool Co. in 1927.

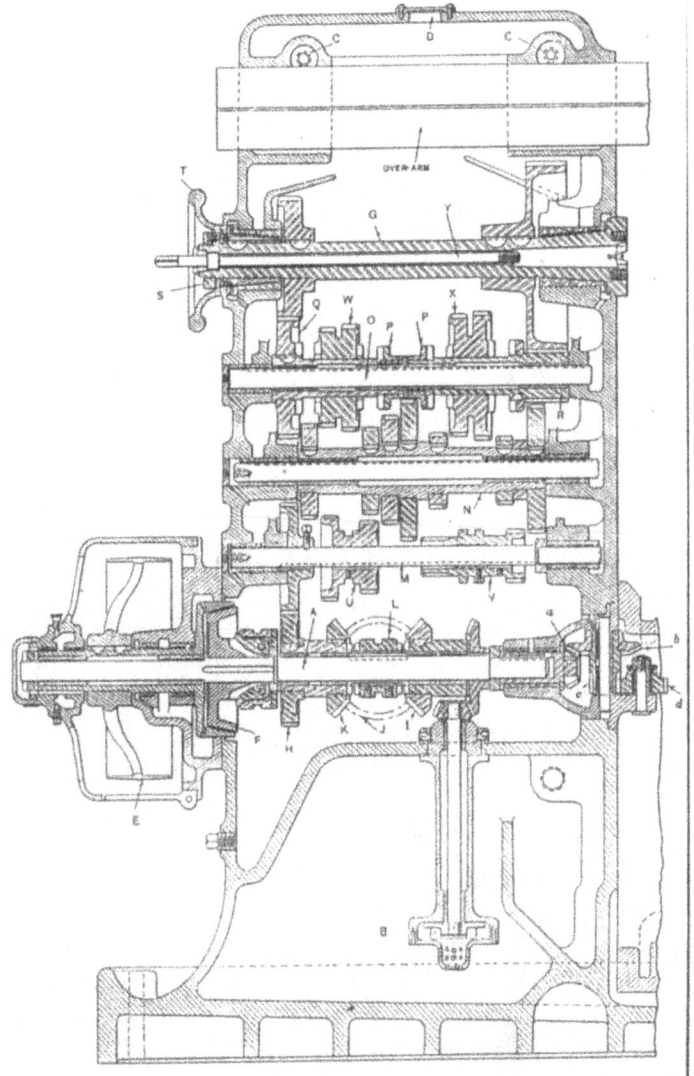

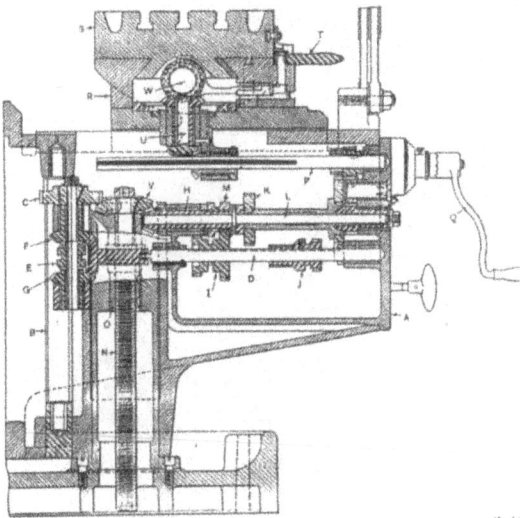

CLOUGH, R.M., Meriden, CT

Founded in 1889 by Roger M. Clough (?-1934) to make gear cutting tools. In 1892, Clough introduced a combination light drill press and milling machine (Fig.1) with 12" table movement, 6" saddle movement and 18" knee movement. Beginning in 1899, the basic machine, coupled with a slotting attachment (Fig.2), was offered as a combined milling, drilling and slotting machine. By 1920, the machine, now minus the slotting function (Fig.3), was offered in a bench version and with spindle speeds up to 3000 RPM. The firm was in business as late as 1935.

(illustrations on next page)

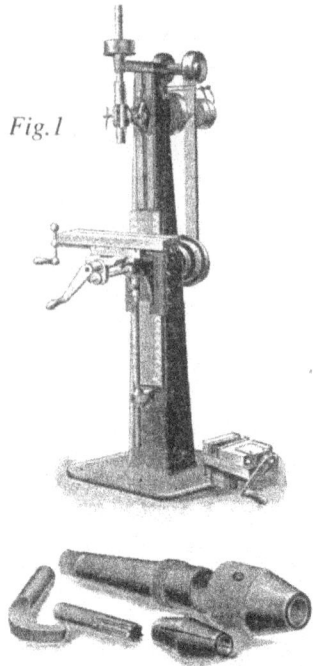

Fig.1

Drill Press and Milling Machine

Fig.2

Fig.3

COCHRANE-BLY MACHINE WORKS, Rochester, NY,

later COCHRANE-BLY CO., Rochester, NY

Organized in 1899 by Robert B. Cochrane, Vincent T. Bly and Myron T. Bly with a capital of only $8,000. Cochrane served as president. By 1912, the firm had reorganized as the Cochrane-Bly Co. Early products were die filing machines and vertical shapers. Plain vertical milling machines (Fig.1) were introduced in 1919 and were available with compound tables (Fig.2) or circular tables with dividing attachment (Fig.3).

Fig.1

Fig.2

Fig.3

COLT PATENT FIREARMS MFG. CO.

Hartford, CT

Maker, beginning in 1848 and continuing into the early 20th century, of the Root milling machine, designed by Elisha K. Root (1808-1865), *right*. The machine was the first to be equipped with a cylindrical overarm and was otherwise similar to the later Lincoln milling machine except that the table motion was driven by a worm meshing with a rack cut into the underside of the table.

Colt machine tools were made primarily for internal use but they were advertised for general sale in 1857 and many were sold or leased to other manufacturers during the Civil War.

THE ROOT MILLING MACHINE.

Fig.1

Fig. 1. Conradson No. 3 Plain Milling Machine with Duplex Helical Drive

CONRADSON MACHINE TOOL CO.

Eau Claire, later Green Bay, WI

Formed by Conrad M. Conradson (1861-1940), who was a noted designer of machine tools, primarily turret lathes. In 1917, he was working at the Phoenix Mfg. Co. in Eau Claire, WI. While there he developed a milling machine (Figs.1-3) which he began to manufacturer under his own name. In 1918, Conradson moved the business to Green Bay, WI and formed the Conradson Machine Tool Co. to make milling machines and planers.

All production was consigned to Joseph T. Ryerson & Son, a Chicago dealer that sold the machines as the

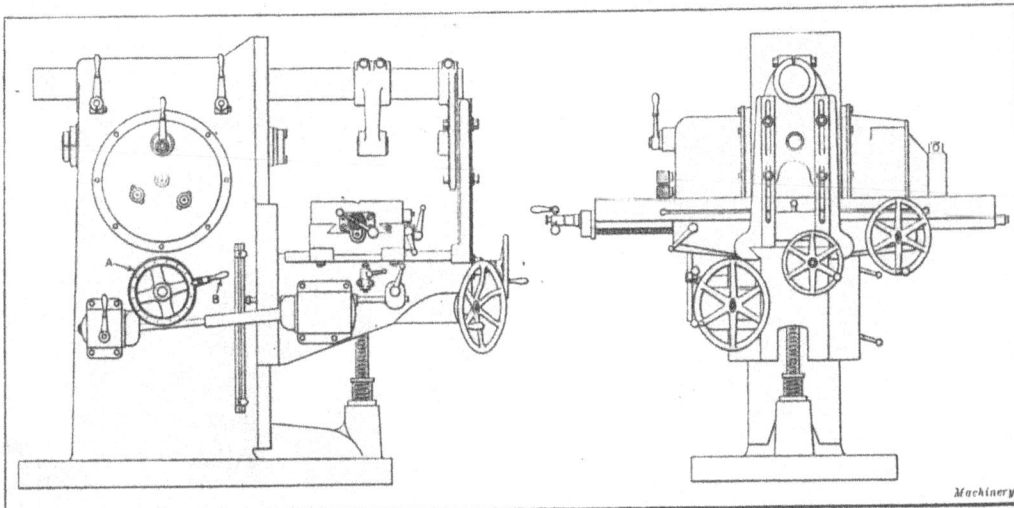

Fig.2

Fig 2 Side and Front Views of Conradson No. 3 Plain Milling Machine shown in Fig. 1

RYERSON-CONRADSON line. 1920 production included a No.3 milling machine (Fig.4) built in plain and universal types, a vertical attachment (Fig.5) and universal dividing head (Fig.6). The new firm failed in 1922. Remaining assets were sold to Joseph T. Ryerson & Son which sold off the property and remaining machines.

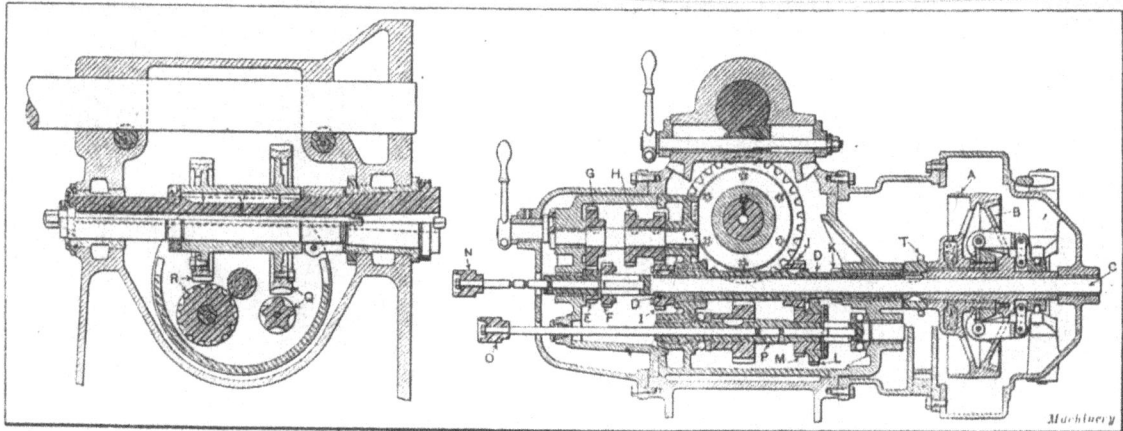

Fig.3

Fig. 3. Cross-sectional View through Speed-box showing Arrangement of Change-gears

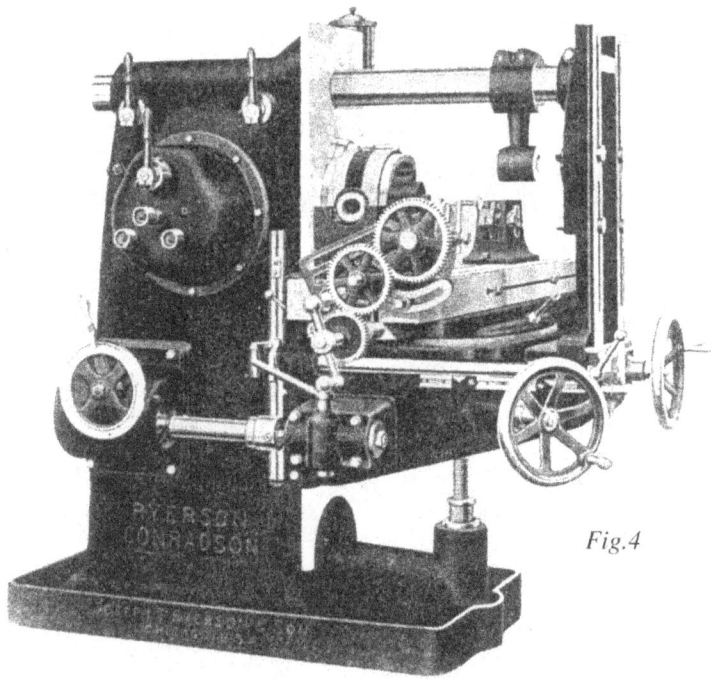

Fig.4

Fig.5

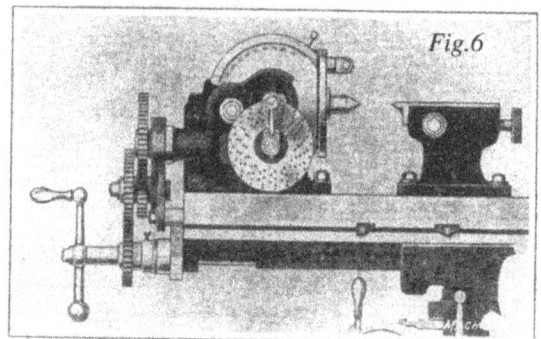

Fig.6

COOMBS & CO., S.C., Worcester, MA

A partnership of Samuel C. Coombs, Russell R. Shepard and Martin Lathe, formed in 1845. Maker of a variety of machine tools, including engine lathes and milling machines. The firm was reorganized as SHEPARD, LATHE & CO. when Coombs left in 1853.

COSTILO MACHINE WORKS, J.E., Brooklyn, NY

Maker, 1899-1900, of hand milling machines, tapping machines, and screw machines.

COX & PRENTISS, Cleveland, OH

A partnership of Jacob D. Cox (1852-1930) and his brother-in-law F.F. Prentiss formed in 1880 when the partnership of NEWTON & COX split into the NEWTON MACHINE TOOL WORKS and Cox & Prentiss. The firm began operating as the Cleveland Twist Drill Co. in 1882 and incorporated under that name in 1904.

1881 milling machine production, some carried over from NEWTON & COX, included improved die-sinking machines (Fig.1) with 16" table travel, 8" cross travel and 17" vertical travel; Lincoln milling machines (Fig.2) weighing 4,200 pounds and claimed to be the largest made in the U.S.; hand milling machines (Fig.3) with 4" table and vertical travel and 6" cross travel. Index centers (Fig.4) and universal vises (Fig.5) were also offered.

Fig.1

IMPROVED DIE-SINKING MACHINE.

Fig.2

LARGE MILLING MACHINE.

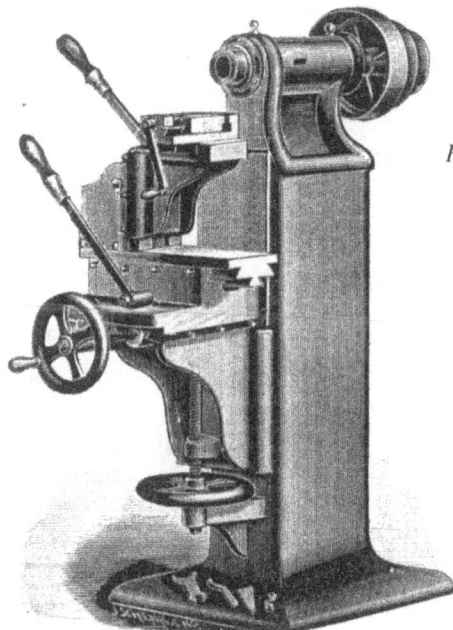

Fig.3

HAND MILLING MACHINE.

Fig.4

INDEX CENTERS.

Fig.5

UNIVERSAL VISE.

CRAFTSMAN TOOL CO., Conneaut, OH

Formed about 1917, the firm offered continuous rotary milling machines for high production of small parts. The machine was furnished in single chuck (Fig.1) and double chuck (Fig.2) models. Production continued until the firm was bought out in 1925.

Fig.1

Fig.2

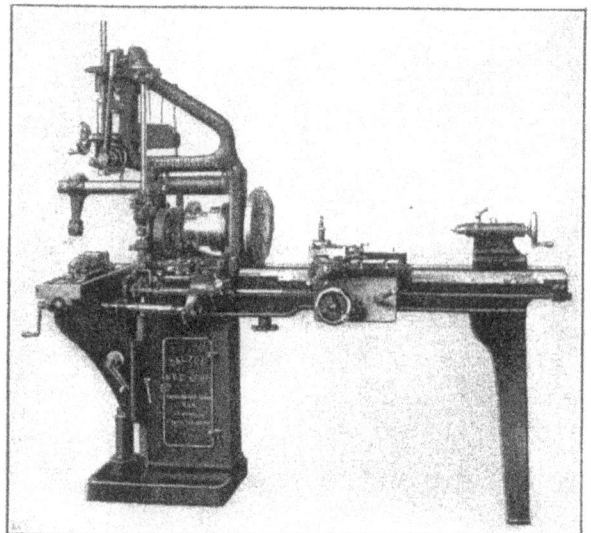

HAND MILLING MACHINE

Adapted for small and duplicate work, built from best materials and with unusual weight and rigidity for its size. Table has long bearing, adjustments can be made without wrenches or other tools and machine is fitted with draw-in spindle and collet or chuck for arbors or end mills.

First class in every particular—send for circular.

DALIN BROS., Rockford, Ill.

DALIN BROS., Rockford, IL

A partnership of Arvid G. Dalin and Gustav A. Dalin (1862-1909) formed in 1897 to make small drilling machines. The firm reorganized as the NATIONAL MACHINE & TOOL WORKS in 1909. In 1908, the firm introduced a small hand milling machine *as shown at left.*

DALTON MFG. CO., Sound Beach, CT

Formed in 1920 as a reorganization of the Dalton Machine Co., a builder of small lathes. Maker of a combination machine "designed primarily for use in garages, on shipboard, and in other places where floor space is at a premium." The machine, *shown at right*, was a combination lathe, milling and drilling machine. The milling machine had table travel of 15", cross travel of 6¾" and vertical travel of 13". In 1924, the firm reorganized as the Dalton Tool Corp.

DANCEL, C., New York, NY

Maker, in 1884, of upright milling machines designed for die sinking work. The machine had a vertical travel of 11", table travel 17" and cross travel 14", and was equipped with a rotary table. Dancel also offered a gear and spiral cutting attachment for the machine.

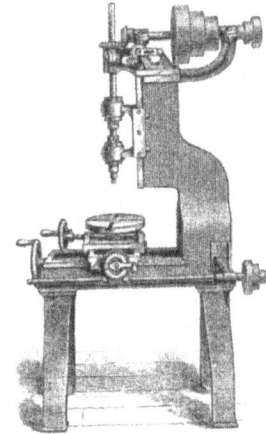

Dancel Vertical Milling Machine.

DAVENPORT MFG. CO., Meadville, PA

Maker, beginning about 1917, of light milling machines (Fig.1) with power or hand feed. A heavier 2-B model (Fig.2) was introduced in 1920.

Built to Meet Your Needs
Fig.1

This heavy duty hand milling machine has been designed with special regard for the needs of the manufacturer who uses this type of machine. It is sturdy, powerful and accurately made throughout. Equipped with power or hand feed.

Send for complete specifications. You'll be interested.

The Davenport Mfg. Co
Meadville, Pa.

A Miller of Great Accuracy

For a milling machine that's rigid, accurate, handy and powerful, there is nothing superior to the—

Davenport No. 2-B

The original construction of arbor support insures long life without wear, prevents chatter and makes possible the finest accuracy on parts requiring exact spacing and thickness. Quick change gear box provides 12 speeds.

Write for specifications, showing the unusual design in detail

THE DAVENPORT MFG. CO., Meadville, Pa.
Fig.2

DAVIS & EGAN MACHINE TOOL CO.

Cincinnati, OH

A partnership of Charles Davis (1853-1903) and Thomas P. Egan, formed September 1, 1896, as a reorganization of the LODGE & DAVIS MACHINE TOOL CO. The company was reorganized in 1898 as the AMERICAN TOOL WORKS CO. when Davis sold out to Franklin Alter. Production of some of the Lodge & Davis Machine Tool Co. milling machines were continued until about 1900. These machines included sprocket milling machines (Fig.1) made for the bicycle industry; Lincoln milling machines (Fig.2) with 18" feed; and universal milling machines (Fig.3) with an improved power feed mechanism.

(illustrations continued on next page)

Fig.1

THE DAVIS & EGAN SPROCKET MILLING MACHINE

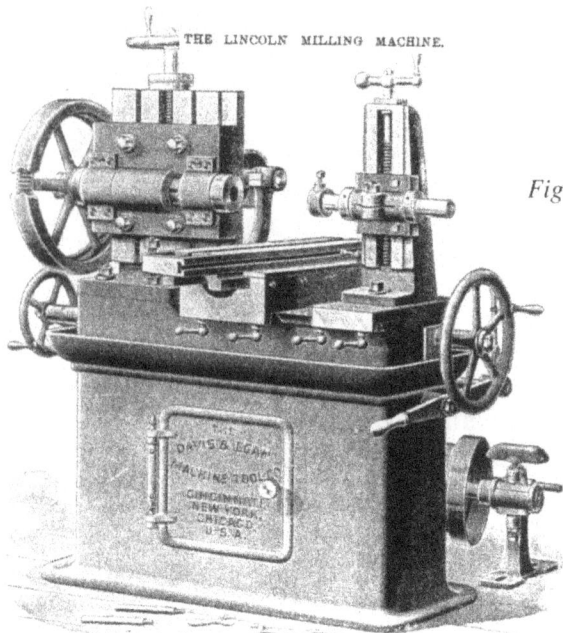

THE LINCOLN MILLING MACHINE.

Fig.2

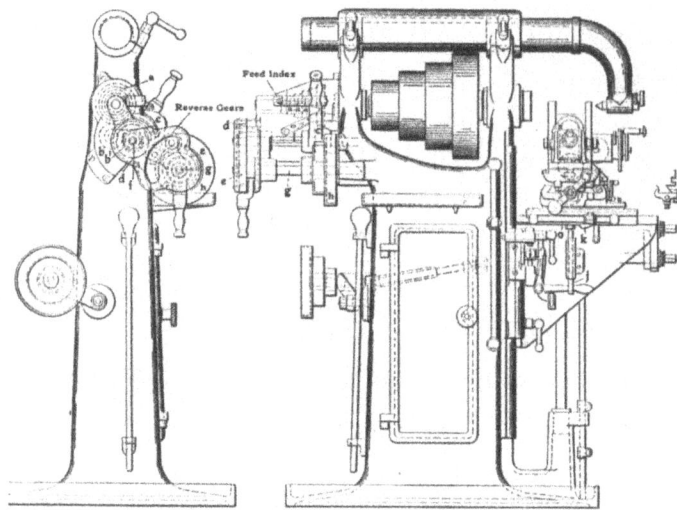

A UNIVERSAL MILLING MACHINE.

Fig.3

DEMCO MACHINE TOOL CO., Cleveland, OH

Formed in 1918 as a reorganization of the DeMooy Machine Co. Products included a combination drilling, milling and tapping machine "Three Machines In One", *as shown at right.*

Three Machines In One
The "Demco" No. 40

Drilling, milling and tapping machine makes a most valuable addition to the labor saving machinery in any shop. It is especially adapted to and is very useful for the making of jigs, dies, tools, etc., where accuracy is absolutely necessary.

There is a very obvious advantage in a single machine which performs all the operations which the "Demco" No. 40 does. It has a reverse drilling capacity of ⅝" and tapping capacity of ⅝". Has four back-geared speeds, right or left, and both positive and sensitive feed.

Write us today for Circulars

The Demco Machine Tool Co.
706 Frankfort St., N.W. Cleveland, O.

Dale-Browater Mchy. Co., New York, Chicago
E. A. Kinsey Co., Indianapolis, Ind.
Radnor-Packard Mchy. Co., Milwaukee, Wis.
Monarch Machinery Co., Philadelphia, Pa.
Wm. C. Johnson & Sons Mchy. Co., St. Louis
W. M. Pattison Supply Co., Detroit
Berger & Carter Co., Los Angeles and San
Francisco, Cal.
The Walraven Co., Atlanta, Ga.

MILLING MACHINE VISE.

Fig.2

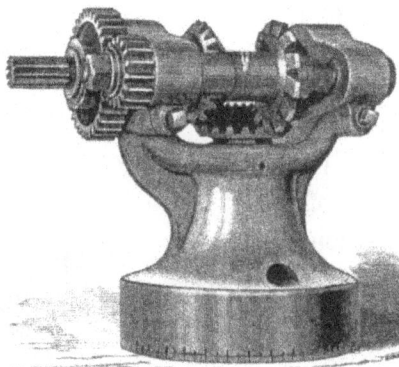

VERTICAL ANGULAR ATTACHMENT.

Fig.3

DETRICK & HARVEY,
Baltimore, MD

A partnership of Jacob S. Detrick and Alexander Harvey formed in 1884 to make open side planers designed by Detrick. In 1885 the firm introduced a heavy milling machine based on Cosgrove's patent of August 7, 1883. The machine (Fig.1 *next page*) was equipped with a 54" x 14" table, rotating on a 16" swivel base, and had 40" table travel, 20" vertical travel, and 12" cross travel. Cosgrove patent milling machine vises (Fig.2) and vertical angular attachments (Fig.3) were offered with the machine. By 1887, the machine was a product of PEDRICK & AYER. *(Detrick Fig.1 at right)*

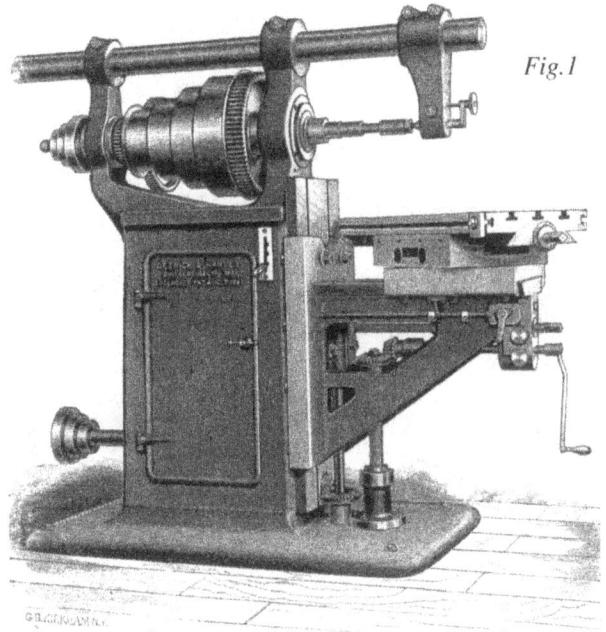

Fig.1

NEW HEAVY MILLING MACHINE.

DICKOW, FRED C., Chicago, IL

Founded in 1905, the firm made the Dickow design index centers (dividing heads) until 1922 or later.

Accuracy Helps Production

DICKOW'S 10-in. Universal Index Centers are built as accurate as human skill can make them, and their dependability in this respect is bound to save you labor and time.

Let us tell you about the improvements.

PROMPT DELIVERY

Patents Pending

For Dependable Accuracy Get Dickow's

Sold by dealers. Write us for complete description.

Fred. C. Dickow, 1858 W. Lake St., Chicago, Ill.

DOW MFG. CO., Braintree, MA

Maker, in 1916-1918, of the plain milling machine *shown at right*. The machine had 21" of table travel, 8" cross travel, and 16" vertical travel. Weight was 2200 pounds.

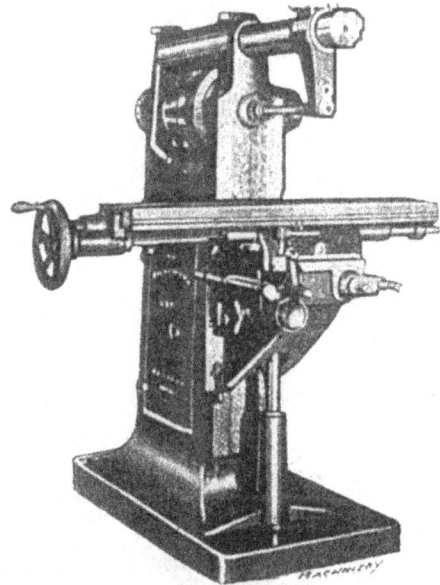

No. 1 Plain Milling Machine built by Dow Mfg. Co.

ELGIN TOOL WORKS, Elgin, IL

Formed by Albert A. Hasselquist (1863-1939) in 1904 as a reorganization of A. HASSELQUIST & CO. The firm continued to make a line of small bench lathes and other small machines for the watchmaking industry. Two types of milling attachments were offered in 1912; a horizontal spindle type (Fig.1) and a vertical spindle type (Fig.2). *(illustrations on next page)*

Elgin Tool

Fig.1

Fig.2

EXCELSIOR MACHINE TOOL CO., Miamisburg, OH

Formed in 1910 as a reorganization of the AL-ED MFG. CO. The firm continued production of hand milling machines, small drilling machines and hand punches previously made by Al-Ed.

EYNON & INGERSOLL, Cleveland, OH

A partnership of William M. Eynon and Winthrop Ingersoll (1865-1928) formed in 1887 to continue the business of W.R. Eynon & Sons. 1887 production included slab milling machines (Fig.1) designed by Ingersoll and offered in several sizes from 12" square to 24" square. Inserted tooth milling cutters (Fig.2), patented December 24, 1889, were offered with the machines. Ingersoll bought control in 1890, reorganized as the INGERSOLL MILLING MACHINE CO., and moved it to Rockford, IL, in April, 1891.

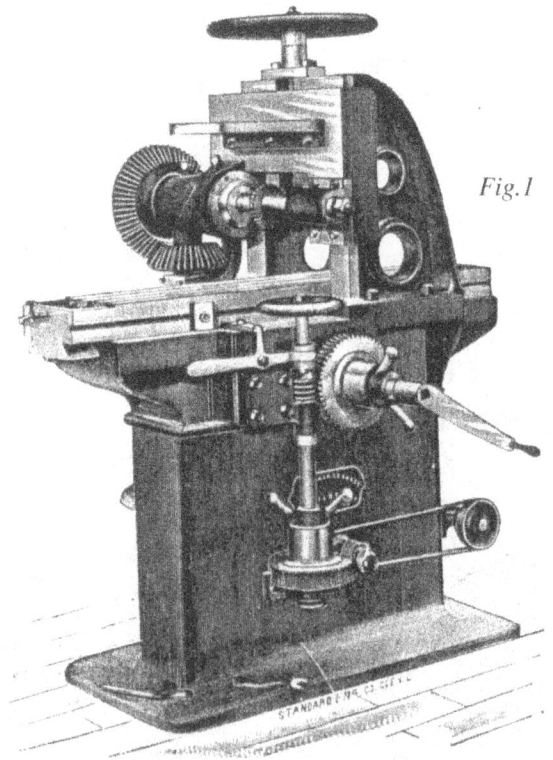

Fig.1

NEW HEAVY MILLING MACHINE.

Pat. Dec. 24, 1889.

Fig.2

FALES, JENKS & SONS, Central Falls, RI

Formed in 1854 as a reorganization of Fales & Jenks, the firm was primarily a pump manufacturer. However, it also made milling machines for the firearms industry during the Civil War.

FALKENAU, A., Philadelphia, PA

Formed in 1891 by Arthur Falkenau. Early products included tool grinders, traveling cranes and engine lathes. In 1898, the firm offered milling machines, shown below, which had been originally developed for its own use. The business continued until 1902 when Falkenau merged with the Philadelphia Machine Tool Co. and reorganized as the FALKENAU-SINCLAIR MACHINE CO.

FANUIEL WATCH TOOL CO., Boston, MA

Founded in 1884 by Edward Rivett (1851-1937) to make small bench machinery for the watch industry. Rivett patented a vertical milling machine November 29, 1892, but it is not clear that it was ever produced. The firm was reorganized as the RIVETT LATHE MFG. CO. about 1900. Production of small bench millers began just before that time.

FAY, L.D., Worcester, MA

Maker of engine lathes and FAY'S combined engine lathe and milling machine, beginning in 1864 and continuing until 1869 or later.

FISH MACHINE WORKS, H.C., Worcester, MA

Formed in 1889 by Henry C. Fish (1825-1906) who had been associated with the L.W. POND MACHINE CO. Early production centered on engine lathes. Keyway milling machines, *shown at right*, were introduced in 1901. Weighing 250 pounds, the machine had a hand feed of 6".

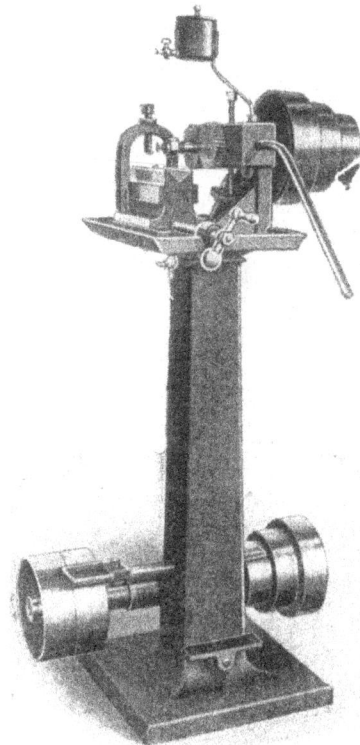

FISHKILL LANDING MACHINE CO., Fishkill Landing, NY

Primarily a steam engine maker, the firm also made milling machines for the firearms industry during the Civil War.

THE FISH KEYWAY MILLING MACHINE.

FLEMING CO.,GEORGE W. Springfield, MA

Formed about 1916 to make engine lathes. In 1918, the firm introduced a combination lathe, boring, and milling machine, shown below. The milling function had table travel of 20", vertical travel of 15" and cross travel of 7½". It appears to be one of the better designs of combination machines but was probably discontinued when the company reorganized as the Fleming Machine Tool Co. in 1919. *(illustration on next page)*

Fleming Co.

FOOTE GEAR & MACHINE CO., J.B., Chicago, IL

Formed in 1904 when the firm of JAMES & FOOTE was split into D.O. James and J.B. Foote Gear & Machine Co. The latter firm continued production, probably for only a short time, of the universal milling machine previously made by JAMES & FOOTE.

FORBES & CO., W.D., Hoboken, NJ

Founded about 1890 by William D. Forbes (1852-1921) who had previously been a partner in Forbes & Curtis, a maker of pipe threading machines. In 1895, Forbes introduced a plain milling machine, with or without back gears, made with a cylindrical column. Forbes claimed the cylindrical design gave "great stiffness." While it is technically correct that a cylindrical form is stiffer than a box form it was probably not noticeable in actual practice.

Plain Miller

with and without Back Gear.

CYLINDRICAL DESIGN.

GIVES GREAT STIFFNESS.

Consequently Extreme Accuracy is Maintained

W. D. FORBES & CO.,

1302 HUDSON ST.,

Hoboken, N. J.

Two blocks from 14th St. Ferry.

FOX MACHINE CO., Grand Rapids, MI, later Jackson, MI

Founded in 1880 by William R. Fox (1853-1934) to make woodworking machinery. In 1892 the firm incorporated with Fox as president and capital of $150,000. Fox moved the company to Jackson, MI in 1916 where he continued to serve as president. Machine tool production began in 1891 when Fox introduced his patented shaper. Milling machine production began in 1893 with the introduction of a two-spindle machine (Fig.1) with 44" table travel and 18" cross travel. A hand feed milling machine (Fig.2) with 9" table travel and 5" cross travel was introduced in 1899. Small milling machine production continued with a combination hand and power feed machine (Fig.3) introduced in 1901, and an improved version (Fig.4) which followed in 1906. The same machine was offered, in slightly improved form (Fig.5), in 1917. The milling machine business was sold in 1920 so that the company could concentrate on drilling machines. *(illustrations on next two pages)*

83

Fig.1

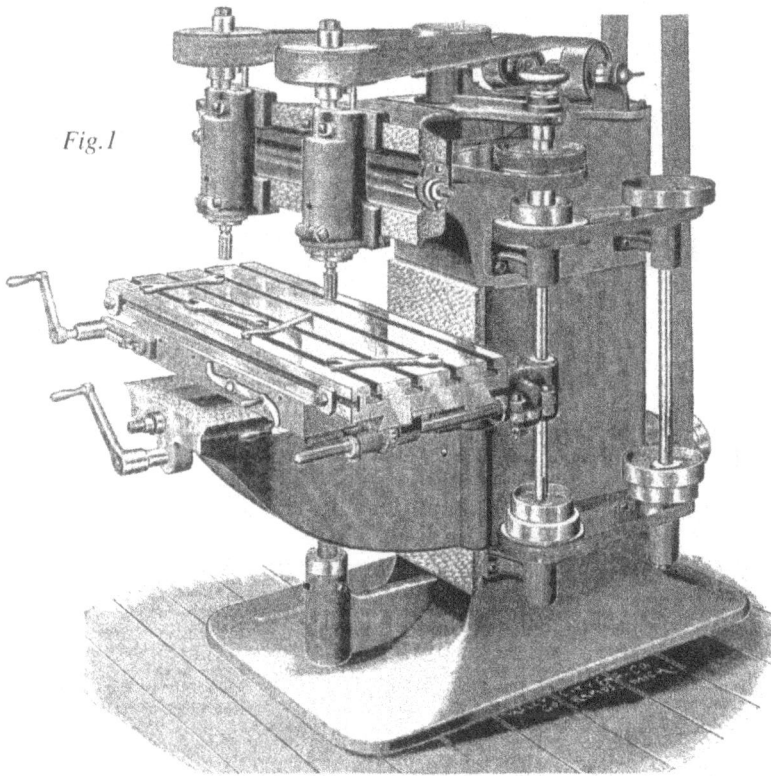

TWO SPINDLE VERTICAL MILLING MACHINE.

NEW HAND-FEED MILLING MACHINE.

Fig 2

Wide Capacity.
Accurate Work.
Perfect Design.

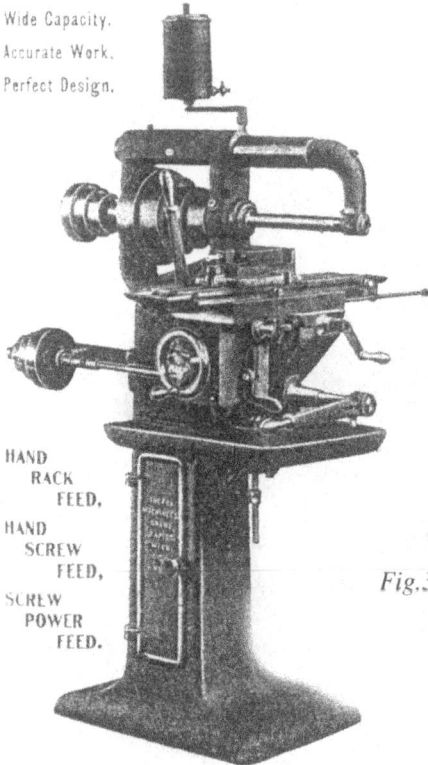

HAND
RACK
FEED,

HAND
SCREW
FEED,

SCREW
POWER
FEED.

Fig.3

WRITE FOR CATALOGUE AND PRICES.

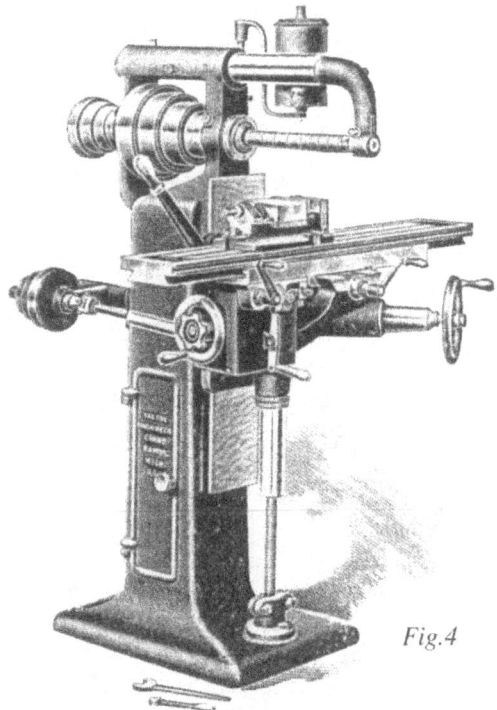

Fig.4

No. 3½ Hand and Power Feed Miller

When the Work Comes
Thick and Fast

Fig.5

You don't have to wonder whether you can "meet" it
or not, if your shop is equipped with a Fox Milling
Machine. They surpass all, in range of work,
accuracy and speed.

**Now is the time to write
for complete details---**

FOX MACHINE CO.

1406 W. Ganson Street
Jackson, Mich.

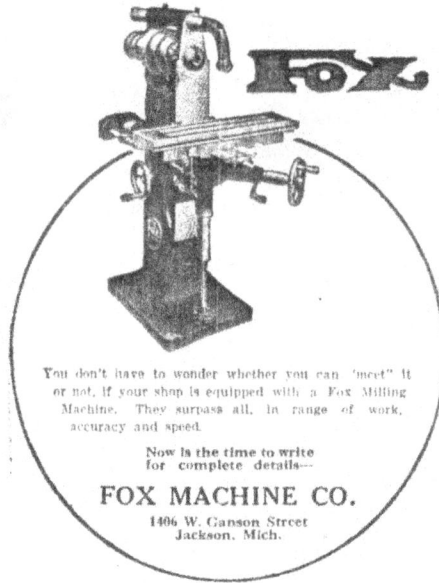

FRANKLIN MACHINE WORKS, Philadelphia, PA

Formed in 1868 to make printing presses. Machine tool production began in 1901 and included heavy milling machines designed by William H. Lucas. The machine, *shown below*, weighed 4000 pounds. Lucas left the firm in 1902 to co-found the Espen-Lucas Machine Co. The Franklin Machine Works failed in 1903.

THE FRANKLIN MILLING MACHINE.

GABRIELSON MFG. CORP., Syracuse, NY

Founded by Carl Gabrielson c1919. Short lived, the firm was reorganized as the Meldrum-Gabrielson Corp. in 1921. Products included a manufacturing milling machine introduced in 1920. It appears to have been made in one size only; specified as 6" size as that was the maximum distance from the arbor to the table top.

Fig. 1. Manufacturing Miller recently developed by the Gabrielson Mfg. Corporation

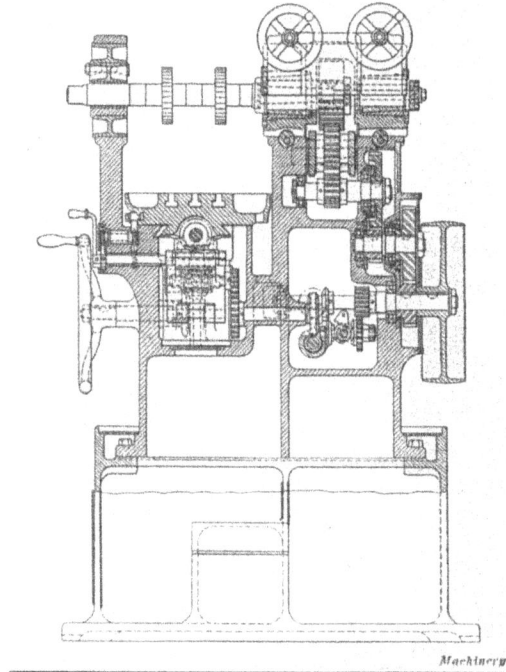

Fig. 2. Cross-sectional View through the Gabrielson Miller showing Arrangement of the Mechanism

GARRIGUS MACHINE CO., C.G.,
Bristol, CT

Formed in 1908, the firm began production of BRISTOL hand feed milling machines in 1913. The machine was equipped with both lever and crank feed and had 18" table travel and 6½" cross travel. In 1918, the firm reorganized as the BRISTOL MACHINE TOOL CO. which does not appear to have continued production of the milling machine line.

Fig. 1—The Operating Side   Fig. 2—The Driving Side

GARVIN & CO., E.E., New York, NY,

later GARVIN MACHINE CO. , New York, NY

A partnership of Eugene E. Garvin (1858-1916) and his brother George K. Garvin (1859-1919) formed in 1879 as a reorganization of SMITH & GARVIN. In mid-1889, the firm became a stock company, the GARVIN MACHINE CO., with George K. Garvin as president and Eugene E. Garvin as vice president. Capital was increased from $100,000 to $500,000 in 1899. Milling machine production in 1880 included small hand milling and screw slotting machines (Fig.1) weighing only 300 pounds; new hand milling machines (Fig.2) with 16" x 4⅝ " tables; profiling machines (Fig.3) with 16" table travel and 9" cross travel; and power milling machines (Fig.4), same as made by Smith & Garvin, with 24" x 6" tables, weighing 1,225 pounds. *(continued on next fourteen pages)*

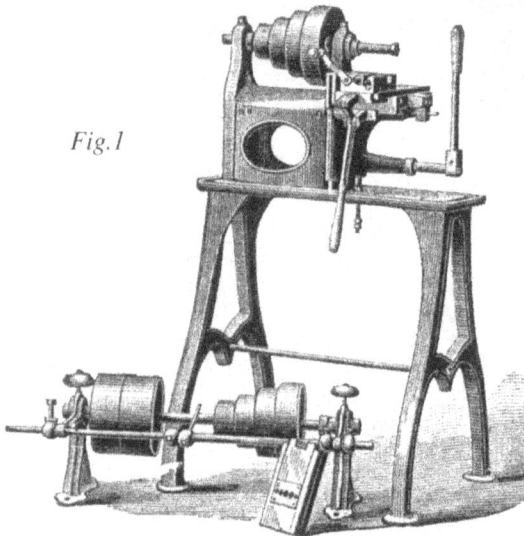

Fig.1

NEW MILLING AND SCREW SLOTTING MACHINE.

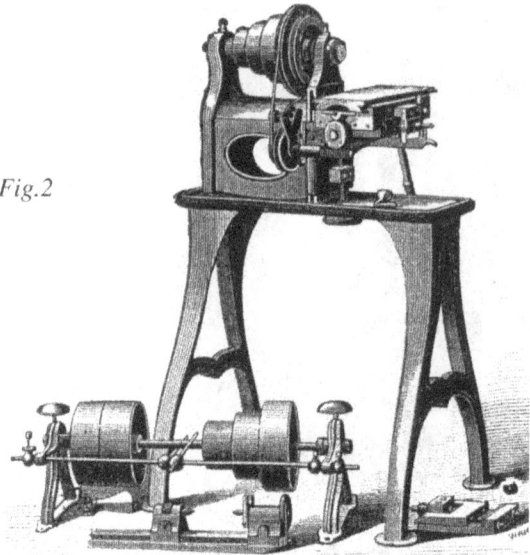

Fig.2

NEW HAND MILLING MACHINE.

PROFILE OR EDGING MACHINE.

Fig.3

POWER MILLING MACHINE.

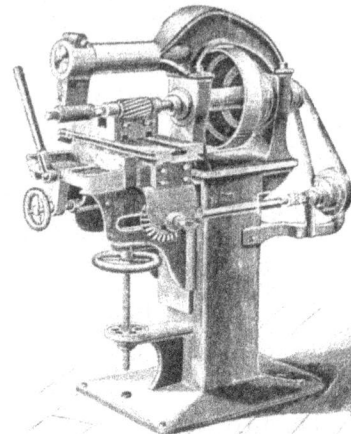

Fig.4

1881 production included new No.2 power milling machines (Fig.5) with 18" table travel and 6" cross travel; new dividing heads (Fig.6) and new spiral attachments (Fig.7), shown mounted on No.2 miller (Fig.8). Back geared milling machines (Fig.9) with 16" table travel and 5" cross travel were introduced in 1882 along with a new screw slotter (Fig.10) weighing 250 pounds. New compound millers (Fig.11) with two heads mounted at 90 degrees, were brought out in 1883.

New products introduced in 1884 included No.5 milling machines (Fig.12), weighing 2,850 pounds; two-spindle milling machines (Fig.(13) for taking rough and finish cuts in one pass, weighing 1,800 pounds; and profiling machines (Fig.14). Other 1884 offerings included small automatic millers (Fig.15); double head face milling machines (Fig.16) weighing 2,500 pounds; smaller power millers (Fig.17) with 24" x 6" tables, weighing 1,150 pounds; and No.3 hand milling machines (Fig.18).

Fig.5

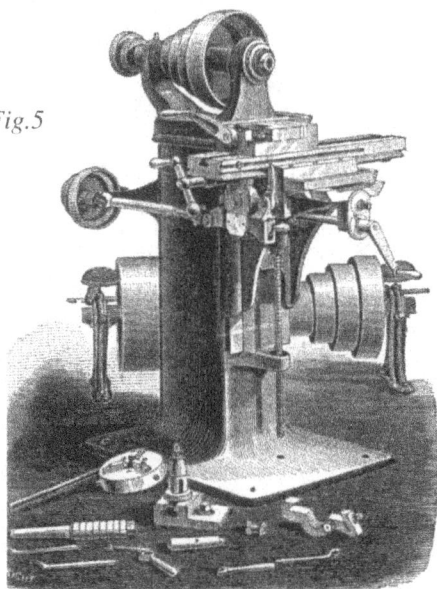

New Power Milling Machine.

Fig.6

DIVIDING HEAD AND TAIL STOCK

Fig.7

Spiral Attachment to Milling Machines.

Fig.8

Spiral Attachment on Milling Machine

Fig.9

Back Geared Milling Machine.

Fig.10

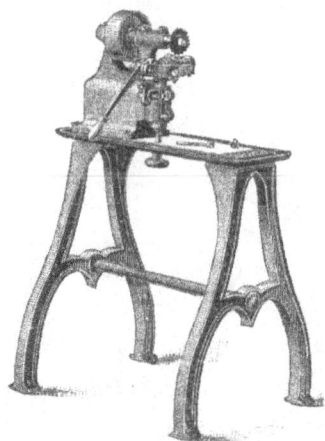

New Screw Slotter.

Fig.11

Compound Milling Machine.

Fig.12

GARVIN'S NO. 5 MILLING MACHINE.

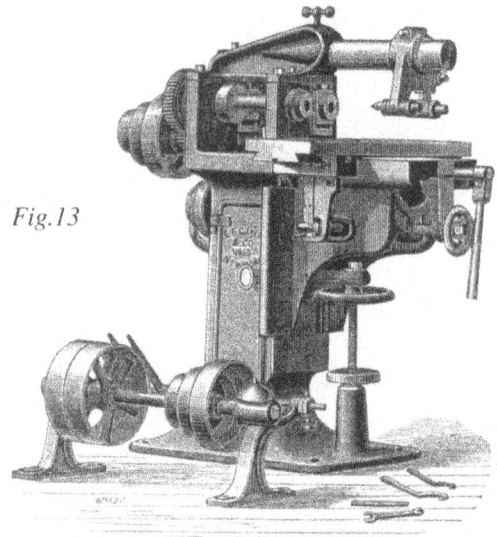

Fig.13

GARVIN'S TWO-SPINDLE MILLING MACHINE.

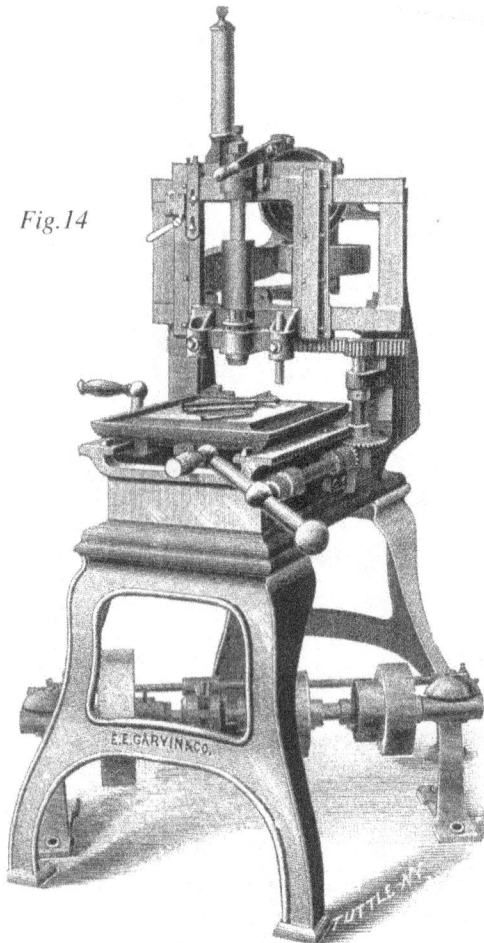

Fig.14

NEW PROFILING MACHINE.

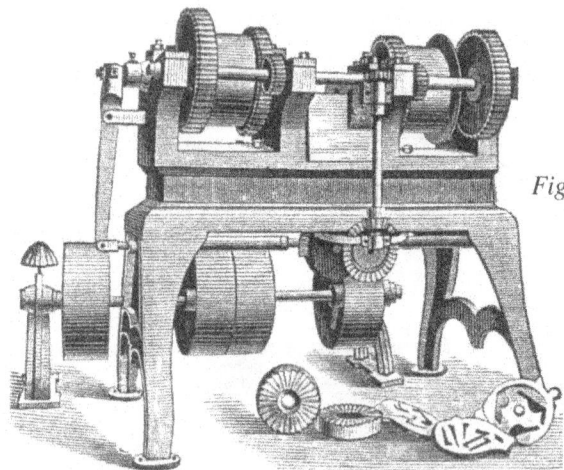

Fig.15

E. E. GARVIN & CO.
139 & 141 CENTRE ST., NEW YORK,

Manufacturers of

MACHINISTS'

TOOLS

Small Automatic Miller,

WITH ARM.

The Machine shown in cut is designed for rapid and convenient milling of small work.

Send for Catalogue.

Fig.16

POWER MILLING MACHINE.

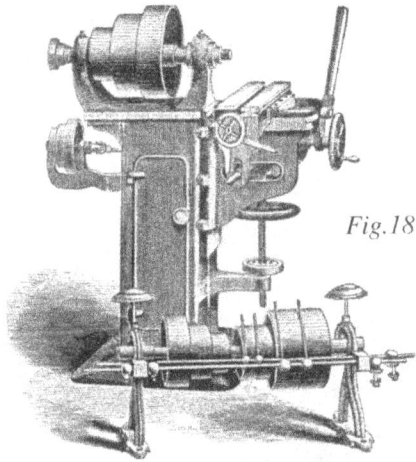

Fig.18

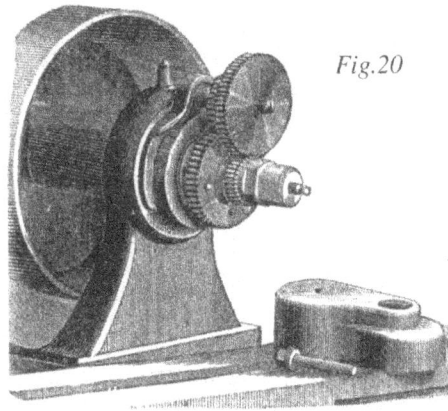

Fig.20

High-Speed Attachment.

Fig.21

Rack-Cutting Attachment.

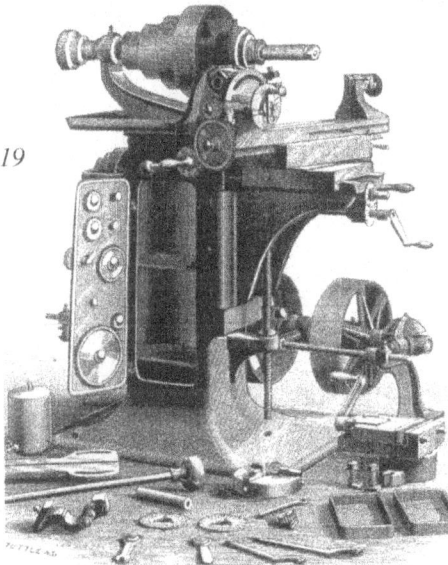

Fig.19

Universal Milling Machine.

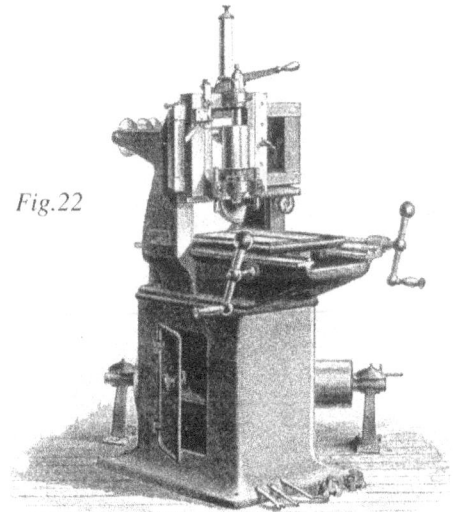

Fig.22

Profiler or Vertical Spindle Milling Machine.

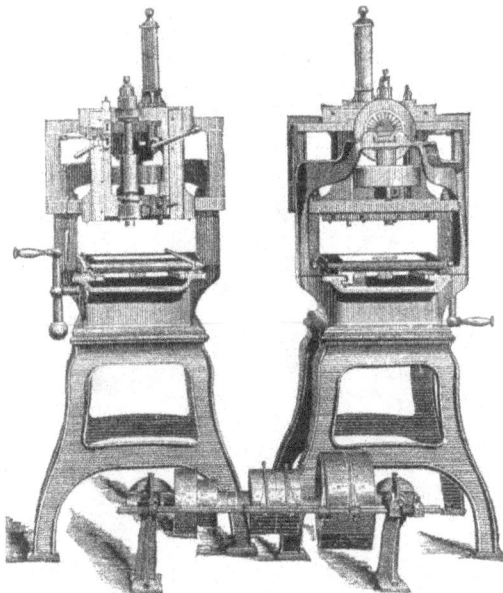

Fig.23

Improved Profile Milling Machine.

Fig.24

Universal Milling Machine.

Fig.25

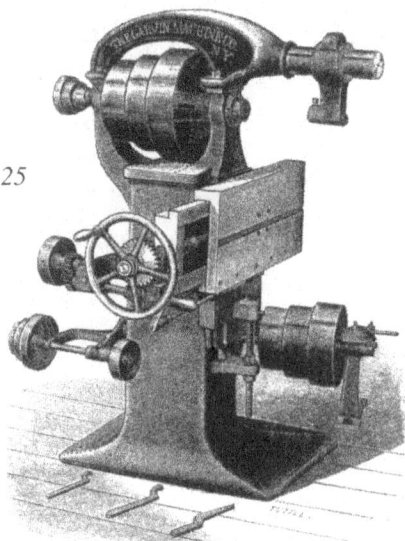

MILLING MACHINE.

Fig.26

UNIVERSAL MILLING MACHINE.

Fig.27

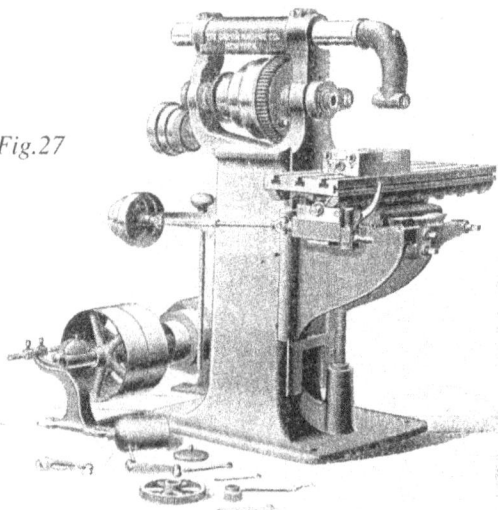

No. 3½ Milling Machine.

Fig.28

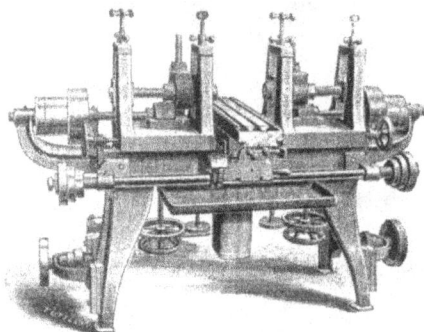

No. 2 Dup'ex Milling Machine.

Fig.29

No. 3 Dup'ex Milling Machine.

Fig.30

UNIVERSAL MILLING MACHINE.

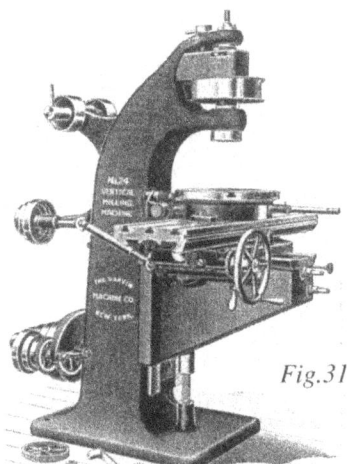

Fig.31

No. 24 Vertical Milling Machine.

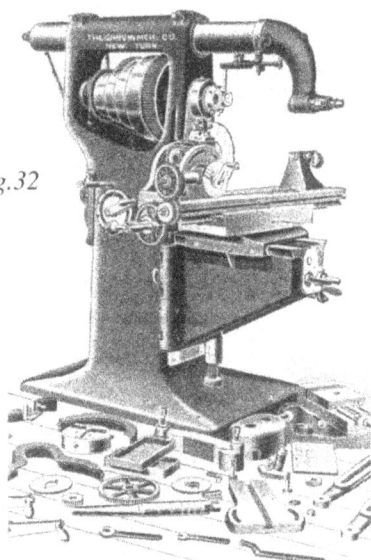

Fig.32

No. 4 Universal

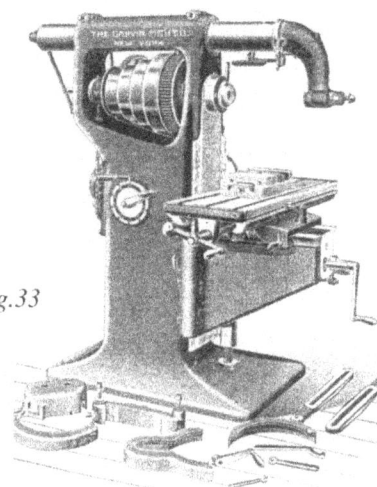

Fig.33

No. 17 Milling Machine.

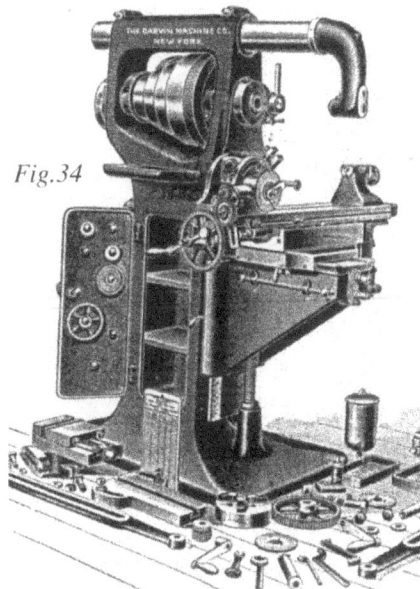

Fig.34

No. 12 MILLING MACHINE.

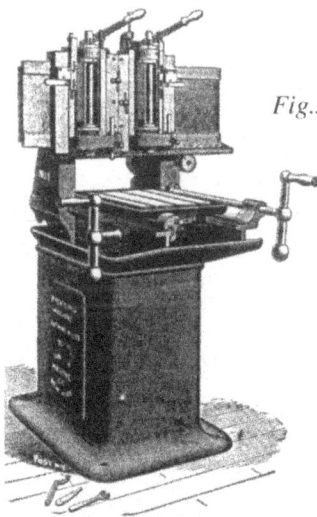

Fig.35

NO. 1 PROFILER.

No. 13 MILLING MACHINE.

Fig.36

NO. 26 VERTICAL MILLER.

No. 14 MILLING MACHINE.

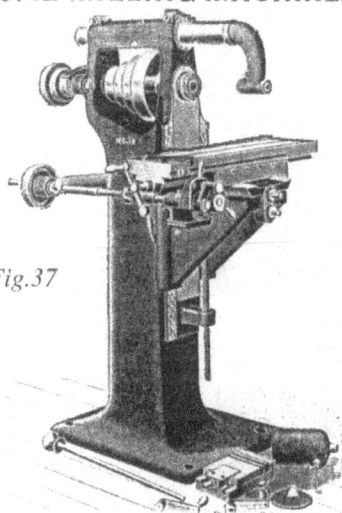

Fig.37

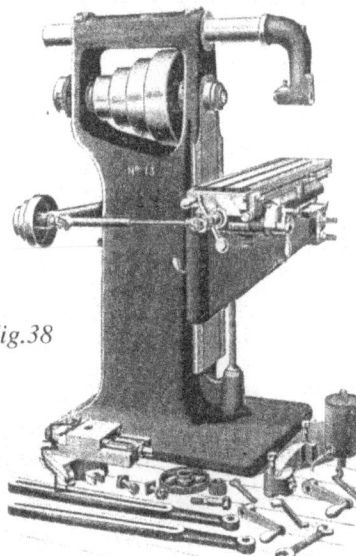

Fig.38

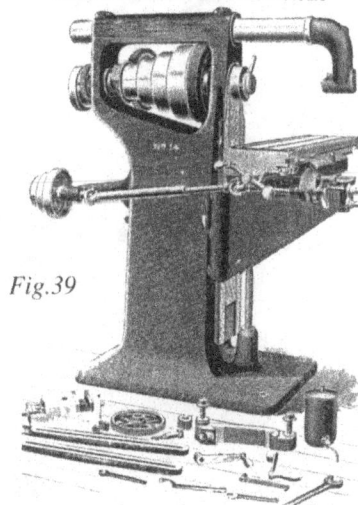

Fig.39

Universal milling machines (Fig.19), weighing 1,350 pounds, were introduced in 1888, along with a high speed attachment for the spindle (Fig.20) and a rack cutting attachment (Fig.21). Other new products in 1888 included improved profilers in two sizes (Figs.22-23); and No.2 universal millers (Fig.24) with overarm.

The Garvin Machine Co., formed by an 1889 reorganization, continued production of the previous machines but soon offered new designs, including milling machines without a horizontal table (Fig.25) designed for mounting fixtures on the moveable platen. A new line of universal milling machines, including the No.3 size (Fig.26) weighing 4,000 pounds, was brought out in 1891; followed, in 1893, by No.3½ plain (Fig.27), No.2 (Fig.28) and No.3 (Fig.29) duplex millers. Another size universal milling machine (Fig.30), weighing 1,950 pounds, was introduced in 1894.

New production in 1898 included No.24 vertical millers (Fig.31); No.4 universal milling machines (Fig.32); and No.17 plain milling machines (Fig.33). No.2 universal milling machines (Fig.34); No.1 profilers (Fig.35); and large No.26 vertical milling machines (Fig.36) were brought out in 1900.

No. 15 PLAIN MILLING MACHINE.

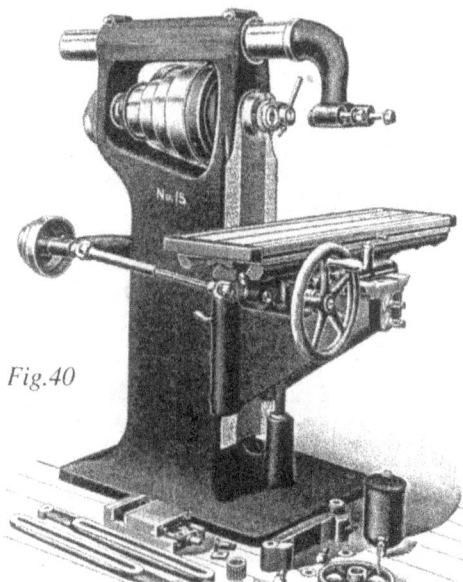

Fig.40

No. 16 PLAIN MILLING MACHINE.

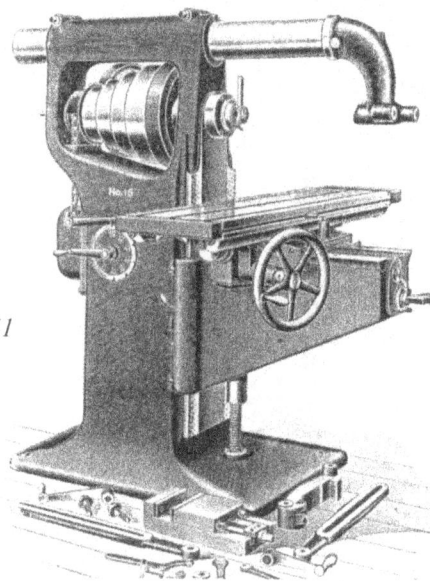

Fig.41

No. 21 Plain Milling Machine.

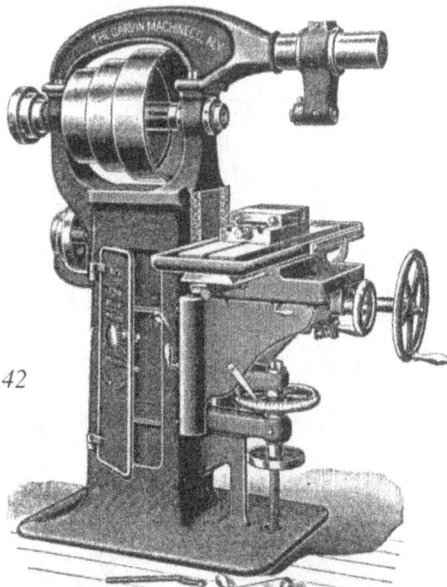

Fig.42

Fig.43

No. 2

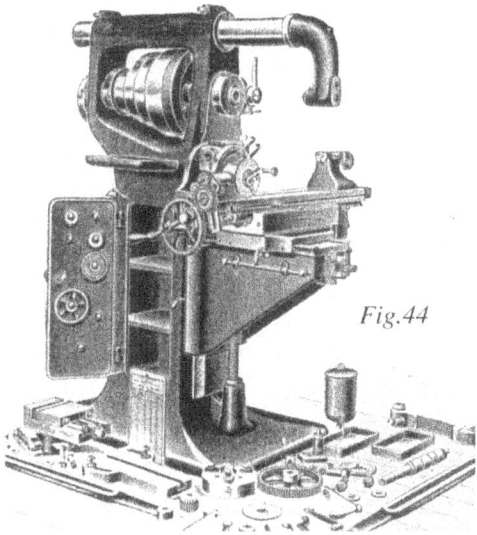

Fig.44

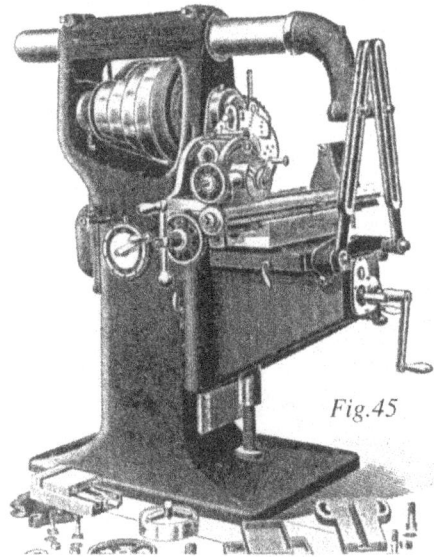

Fig.45

No. 13 Plain Milling Machine.

No. 4 UNIVERSAL MILLING MACHINE.

Fig.46

No. 14 Plain Milling Machine.

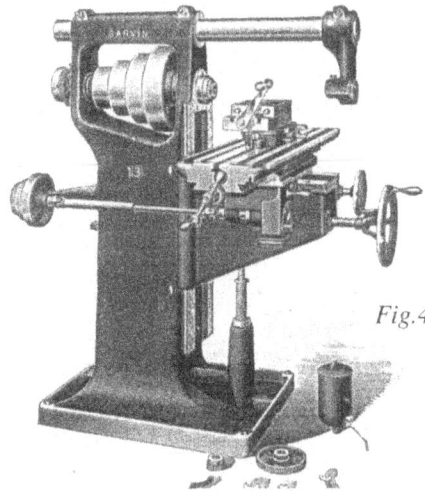

Fig.47

No. 15 Plain Milling Machine.

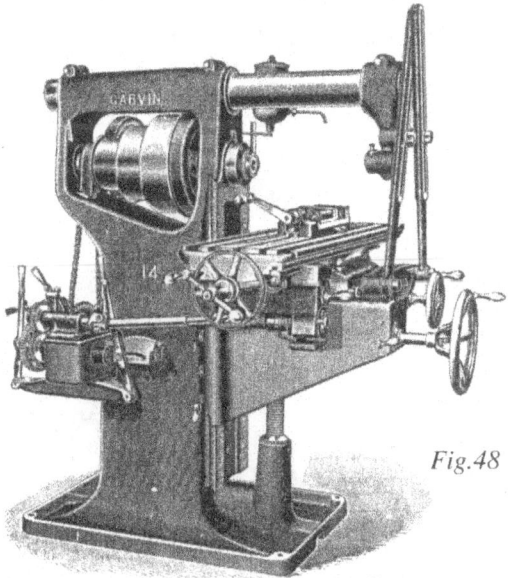

Fig.48

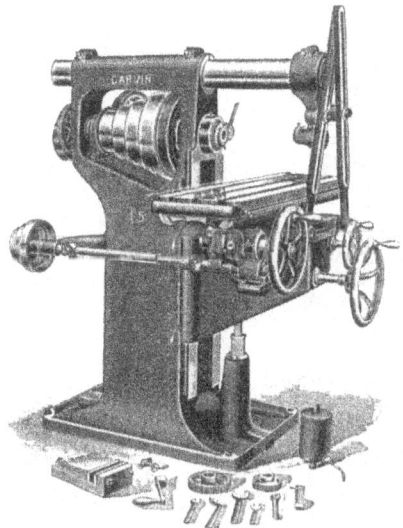

Fig.49

Weight, 2800 lbs.

No. 16A Plain Milling Machine.

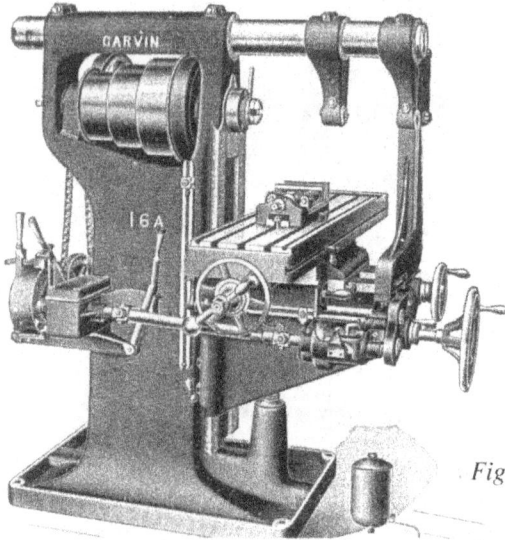

Fig.50

No. 1½ Universal Milling Machine.

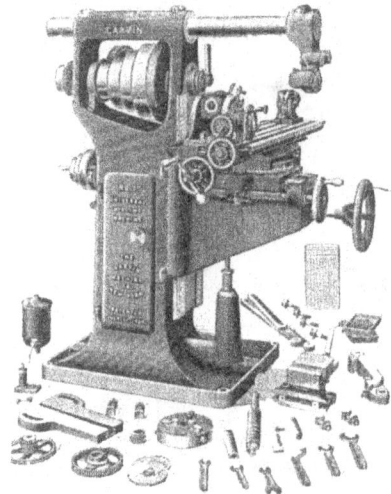

Fig.51

Weight, 2035 lbs.

No. 2A Universal Milling Machine.

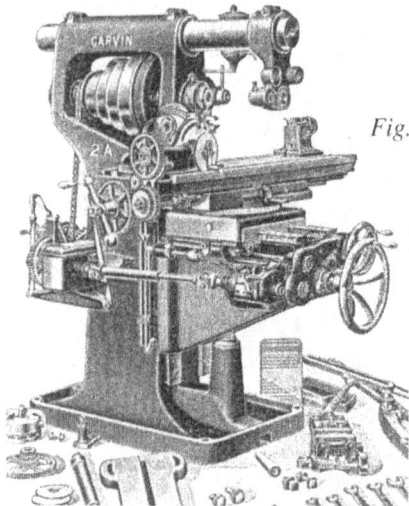

Fig.52

No. 3A Universal Milling Machine

Fig.53

Weight, 4500 lbs.

Fig.54

No. 22 Size.   Weight, 1910 lbs.

Fig.55

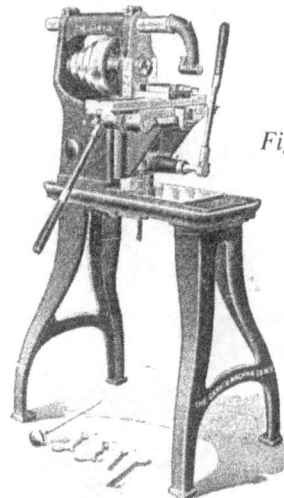

Fig.56

Weight, 631 lbs.

1901 production, made under patents of June 26, 1900, and July 31, 1900, included plain milling machines in No.12 (Fig.37), No.13 (Fig.38), No.14 (Fig.39), No.15 (Fig.40), No.16 (Fig.41) and No.21 (Fig.42) sizes; universal milling machines in No.1 (Fig.43), No.2 (Fig.44), No.3 (Fig.45), and No.4 (Fig.46) sizes.

New or improved milling machines, patented February 9, 1904, were offered in 1906 and included plain milling machines in No.13 (Fig.47), No.14 (Fig.48), No.15 (Fig.49) and No.16 (Fig.50) sizes; universal milling machines in No.1½ (Fig.51), No.2A (Fig.52), and No.3A (Fig.53) sizes; vertical milling machines in No.22 (Fig.54) size; hand milling machines in No.1 (Fig.55), No.2 (Fig.56), and No.3 (Fig.57) sizes; No.9 manufacturing millers (Fig.58); No.1 (Fig.59) and No.2 (Fig.60) lincoln millers; duplex millers in No.1 (Fig.61), No.2 (Fig.62), No.3 (Fig.63) and No.4 (Fig.64) sizes; No.3 (Fig.65) and No.5 (Fig.66) profilers.

No. 3 Hand Milling Machine.

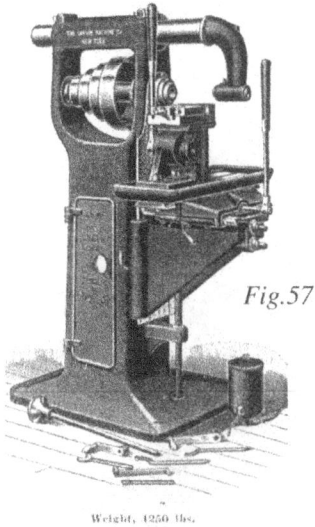

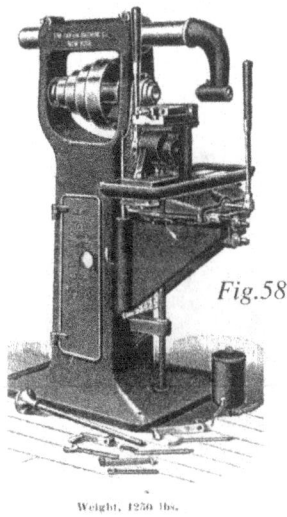

No. 3 Hand Milling Machine.

Fig.57

Fig.58

Weight, 1250 lbs.

Weight, 1250 lbs.

No. 1 Lincoln Milling Machine

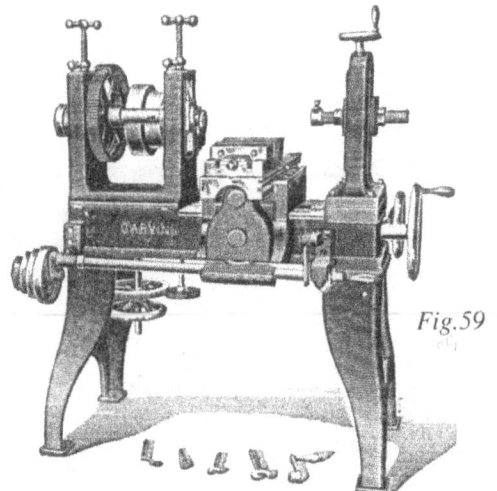

Fig.59

No. 1 Duplex Milling Machine.

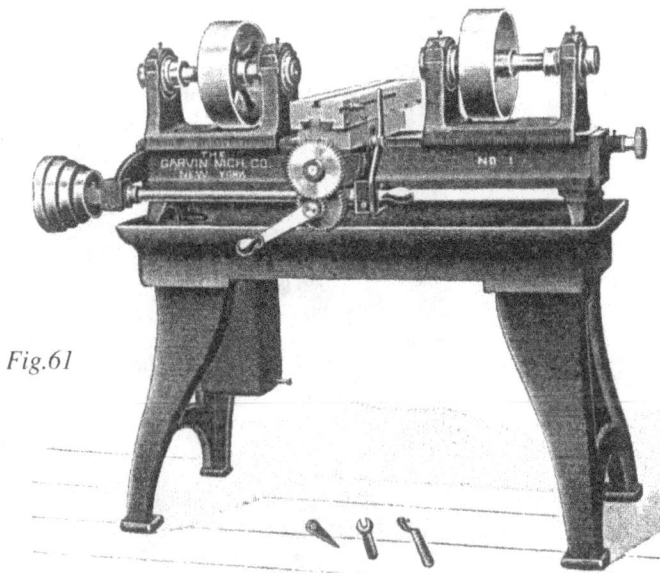

Fig.61

No. 2 Lincoln Milling Machine.

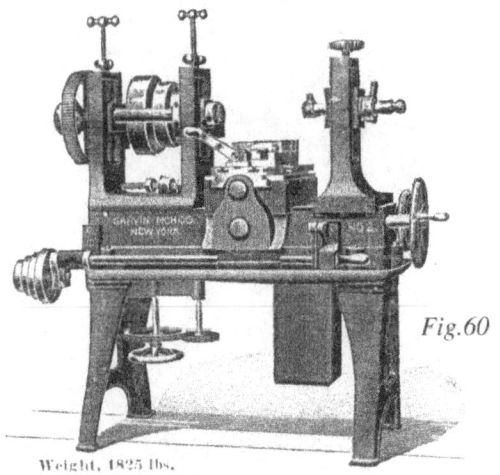

Fig.60

Weight, 1825 lbs.

Weight, 1225 lbs.   Made only to order.

No. 2 Duplex Milling Machine.

Fig.62

Weight, 2465 lbs.   Made only to order.

No. 3 Duplex Milling Machine.

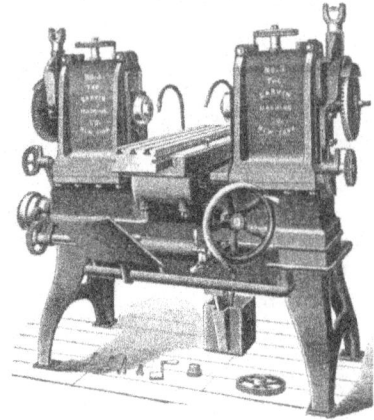

Fig.63

Weight, 3125 lbs.   Made only to order.

No. 4 Duplex Milling Machine.

Fig.64

Weight, 9500 lbs.

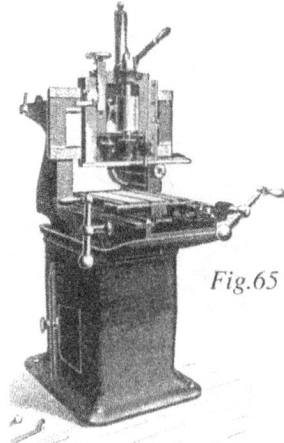

Fig.65

Fig.66

Fig.67

NO. 1 UNIVERSAL MILLING MACHINE, BACK-GEARED

(Former size No. 1½)

Fig.68

**NO. 2 MANUFACTURING UNIVERSAL MILLING
MACHINE**

Garvin's 1919 catalog listings included universal milling machines in No.1 (Fig.67), No.2 (Fig.68), No.2A (Fig.69), and No.3A (Fig.70) sizes; plain milling machines in No.2A (Fig.71), No.3A (Fig.72), No.11 (Fig.73), No.12 (Fig.74), No.13 (Fig.75), No.14 (Fig.76), No.15 (Fig.77), and No.21 (Fig.78) sizes; vertical milling machines in No.20 (Fig.79) and No.22 (Fig.80) sizes; profiling machines in No.0 (Fig.81), No.1 (Fig.82), No.2 Wesson drive (Fig.83), No.2 belt drive (Fig.84), and No.4 (Fig.85) sizes; vertical milling machines in No.14 (Fig.86) and No.15 (Fig.87) sizes; duplex milling machines in No.1 (Fig.88) size; duplex slot milling machines in No.0 (Fig.89), No.1 (Fig.90), and No.2 (Fig.91) sizes; cam milling machines in 6" (Fig.92), 12" (Fig.93), and 24" (Fig.94) sizes; universal milling attachments (Fig.95); vertical spindle attachments (Figs.96-97); and index centers (Figs.98-99).

The company failed in 1925; the various product lines were auctioned off with the milling machine line going to Frew Machine Co. and H. Leach Machinery Co. Neither firm built any machines but no doubt furnished parts for existing machines.

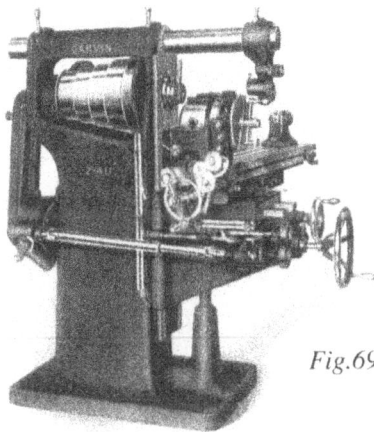

Fig.69

NO. 2-A (ALL FEEDS) UNIVERSAL MILLING MACHINE

Fig.70

NO. 3 A (ALL FEEDS) UNIVERSAL MILLING MACHINE
With Double Back Gears

Fig.71

NO. 2-A (All Feeds) PLAIN MILLING MACHINE

Fig.72

NO. 3-A (All Feeds) PLAIN MILLING MACHINE
With Double Back Gears

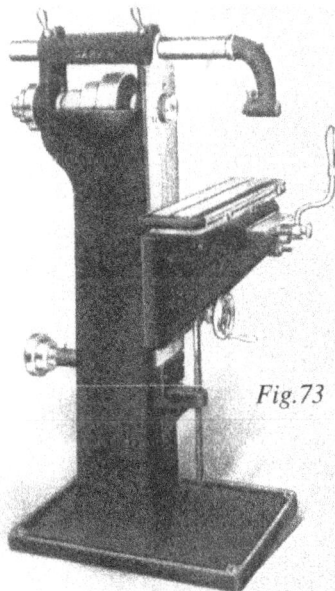

Fig.73

NO. 11 PLAIN MILLING MACHINE

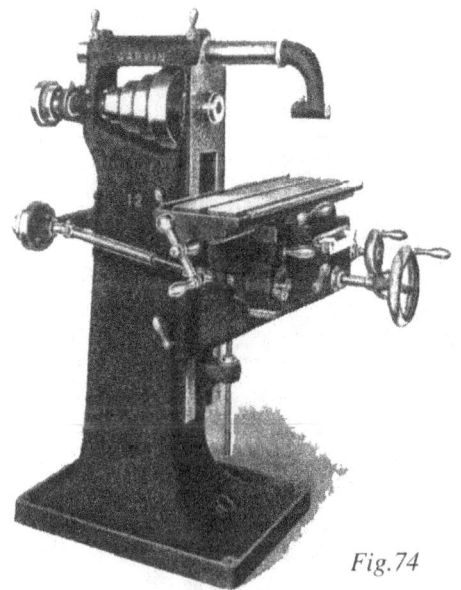

Fig.74

NO. 12 PLAIN MILLING MACHINE

Fig.75

NO. 13 PLAIN MILLING MACHINE

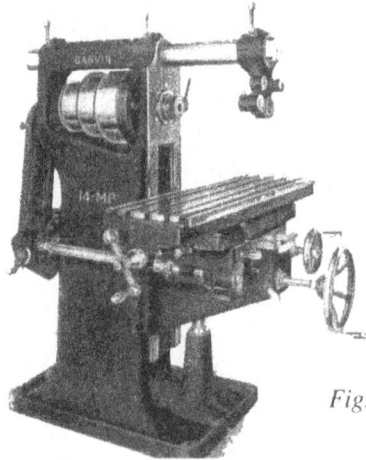

Fig.76

NO. 14 MANUFACTURING PLAIN MILLING MACHINE

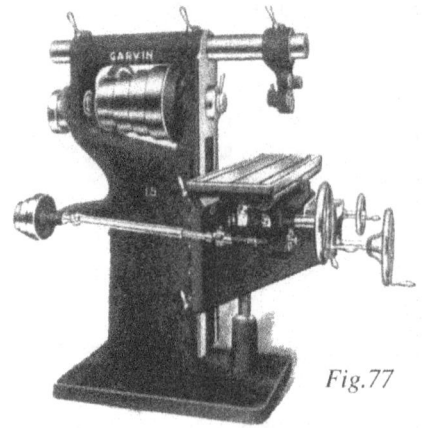

Fig.77

NO. 15 PLAIN MILLING MACHINE

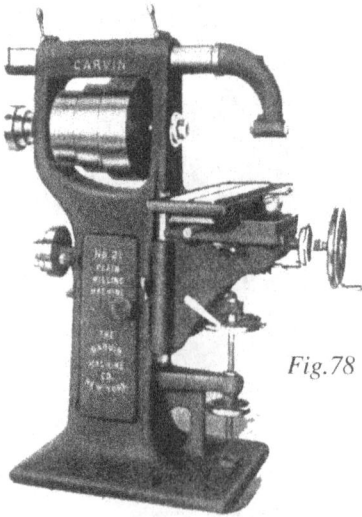

Fig.78

NO. 21 PLAIN MILLING MACHINE

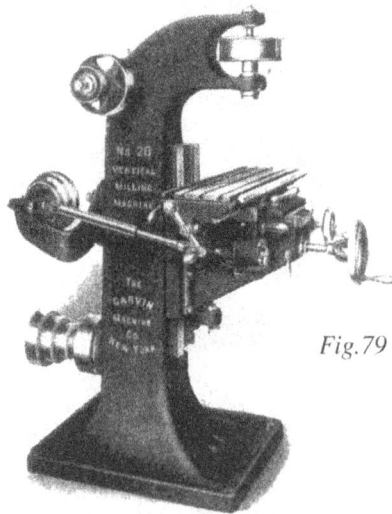

Fig.79

NO. 20 VERTICAL MILLING MACHINE

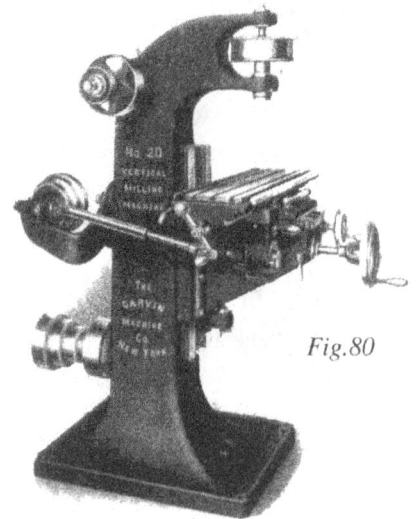

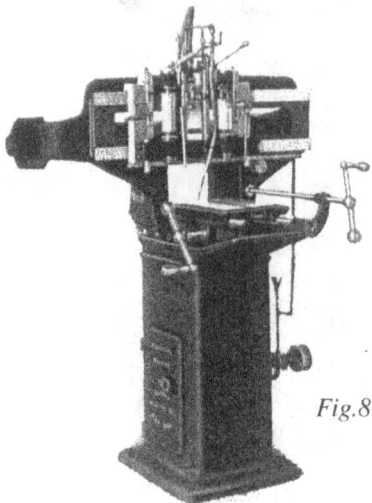

Fig.80

NO. 20 VERTICAL MILLING MACHINE

Fig.81

NO. 0 TWO-SPINDLE PROFILING MACHINE

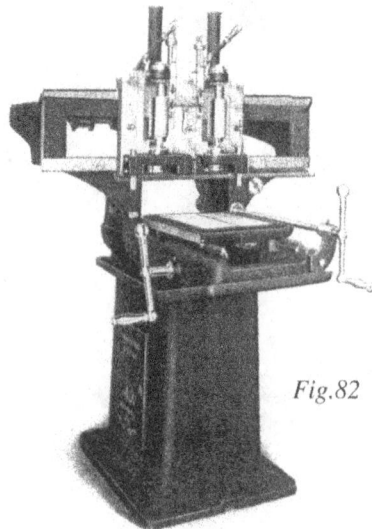

Fig.82

NO. 1 TWO-SPINDLE PROFILING MACHINE

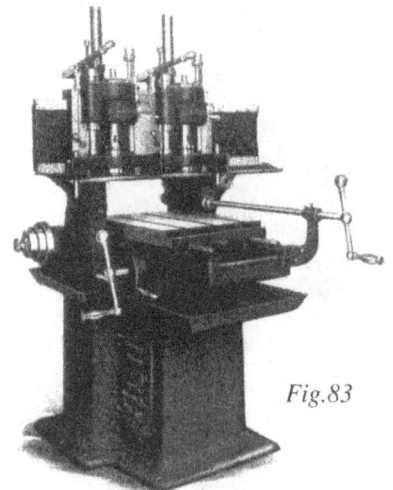

Fig.83

NO. 2 TWO-SPINDLE PROFILING MACHINE

Fig.84

NO. 2 TWO SPINDLE PROFILING MACHINE

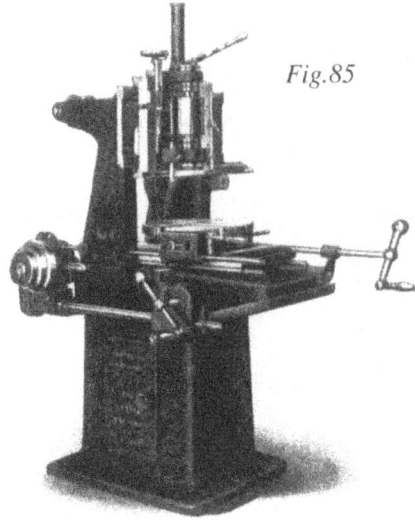

Fig.85

NO. 4 PROFILING MACHINE

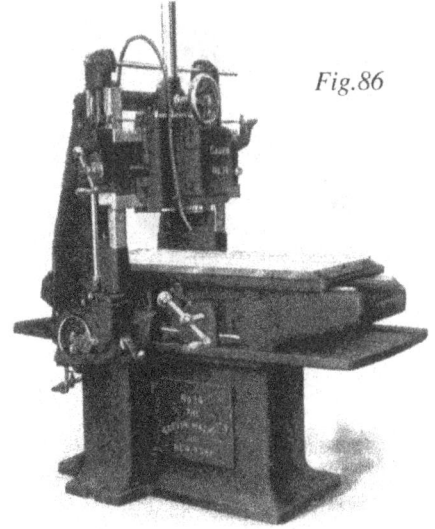

Fig.86

NO. 14 VERTICAL SPINDLE MILLING MACHINE

Fig.87

NO. 15 VERTICAL SPINDLE MILLING MACHINE

Fig.88

NO. 1 DUPLEX MILLING MACHINE

Fig.89

NO. 0 DUPLEX SLOT MILLING MACHINE

Fig.90

NO. 1 DUPLEX SLOT MILLING MACHINE

Fig.91

NO. 2 DUPLEX SLOT MILLING MACHINE

Fig.92

6-INCH CAM OR FORM MILLING MACHINE

Fig.93

12-INCH CAM OR FORM MILLING MACHINE

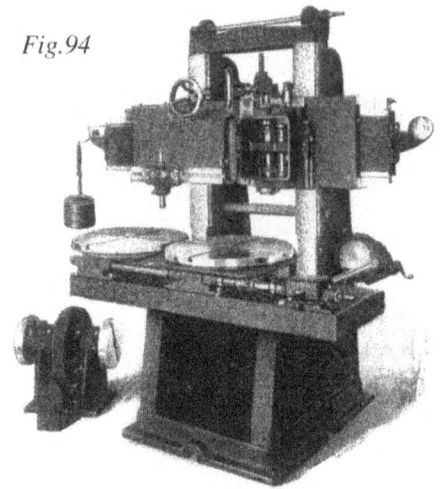

Fig.94

24 AND 36-INCH CAM OR FORM MILLING MACHINES

Fig.95

Nos. 1 and 2

Fig.96

Nos. 000-V, 00-V, 0-V, 1-V and 2-V

Fig.97

No. 3-V

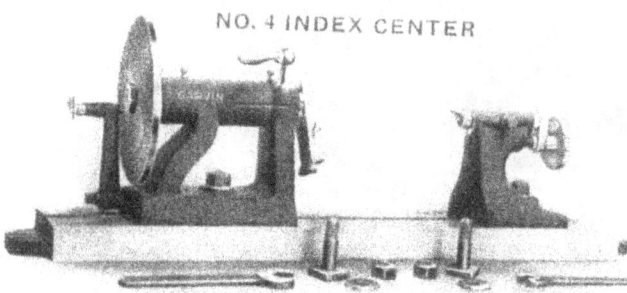

NO. 4 INDEX CENTER

Fig.98

NO. 6 INDEX CENTER

Fig.99

GAY, IRA & ZIBA, North Chelmsford, MA,

later **GAY, SILVER & CO**. North Chelmsford, MA

A partnership of brothers Ira Gay (?-1835) and Ziba Gay (?-1857), formed in 1830 to make textile machinery, lathes and other machine tools. The firm reorganized as Gay, Silver & Co. when Ira Gay died in 1835 and was replaced by Harvey Silver. Frederick W. Howe learned the machinist trade here before joining ROBBINS & LAWRENCE about 1853. The firm was again reorganized, as SILVER, GAY & CO., in 1857 when Ziba Gay died and was replaced by his son Ziba Gay, Jr. (1823-1902). The milling machine, *shown at right*, is believed to have been made by Gay, Silver & Co. in the 1830s and was still in use when photographed in 1924.

GOOLEY & EDLUND, Syracuse, NY

later Cortland, NY

A partnership of Malichi J. Gooley (1845-1928) and John A. Edlund (1866-1931) formed in 1912 to make the BRIGGS milling machine. The firm moved to Cortland, NY, in June, 1914, where it was incorporated as Gooley & Edlund, Inc. Gooley's nephew, Patrick J. Davern, bought out both Gooley and Edlund in 1922 and continued operation until 1931. The Briggs milling machine, *shown at left*, was of the manufacturing type and was offered in table lengths of 32', 38' and 44" with corresponding travels of 16", 22" and 28".

GORTON MACHINE CO., GEORGE, Racine, WI

Founded in 1893 by George Gorton, Jr. Early products were basket making machinery, with machine tool production beginning about 1900. Gorton's first milling machine appears to be a high speed vertical miller (Fig.1) introduced in 1915 for use in making shrapnel shells. The 8-C Universal Horizontal Routing Machine (Fig.2) was offered in 1916 and was designed for milling grooves and vents in artillery shells. Gorton claimed that the machine was available for delivery at the rate of 45 machines per week. The No.1-S Universal Die Sinking and Engraving Machine (Fig.3) was introduced about 1920 for heavy engraving work on embossing dies, forging dies, etc. *(illustrations on next page)*

Fig.1

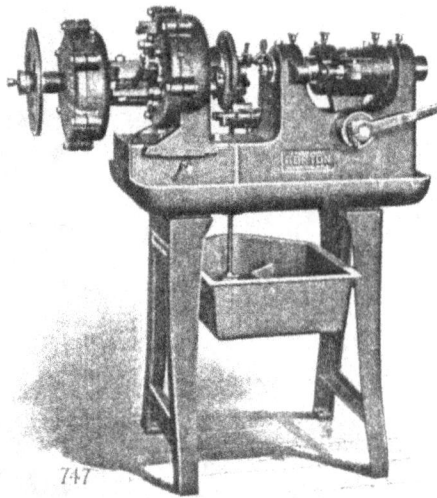

Fig.2

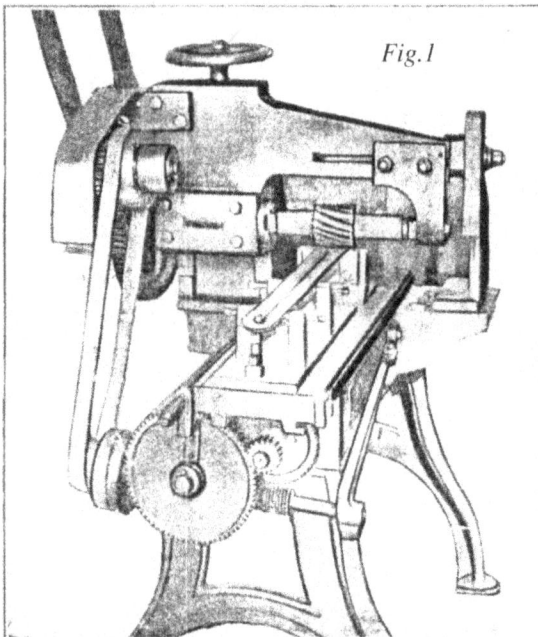

Fig.3

GOULD, E., Newark, NJ

later **GOULD & CO., E.**, Newark, NJ,

later **GOULD MACHINE CO.**, Newark, NJ,

later **GOULD E. & R.J.**, Newark, NJ

Founded in 1833 by Ezra Gould (1808-1901). Gould reorganized as E. Gould & Co. when his brother Stephen Gould joined him in 1857. The firm reorganized as the Gould Machine Co. when his two sons, Frank and Roscoe J. Gould, and Ulrich Eberhardt (1841-1901) became partners in 1862; and again in 1870 as E. & R.J. Gould. Gould's main products were planers, shapers, and woodworking machinery. However, the firm produced a line of milling machines from c1858 (Fig.1) which was improved by c1865 (Fig.2) and again by c1870 (Fig.3). The machines were photographed in 1912, still in production.*(illustration continued next page)*

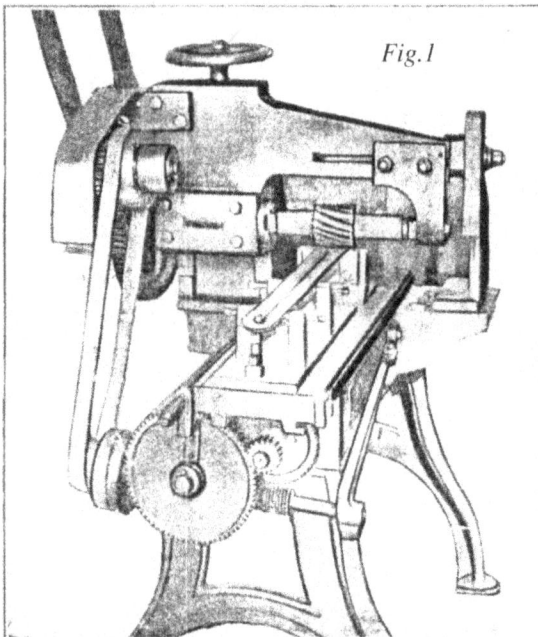

Fig.1

Fig.2

Gould
Fig.3

GRAND RAPIDS MACHINE TOOL CO.,

Grand Rapids, MI

Founded in 1910, the firm was merged into the Valley City Machine Works in 1913. Products included a plain milling machine, introduced in 1910, with a 32" x 8" table, and 24" table travel, 8" cross travel, and 18" vertical travel. Sliding gears gave 12 feed changes. *(right)*

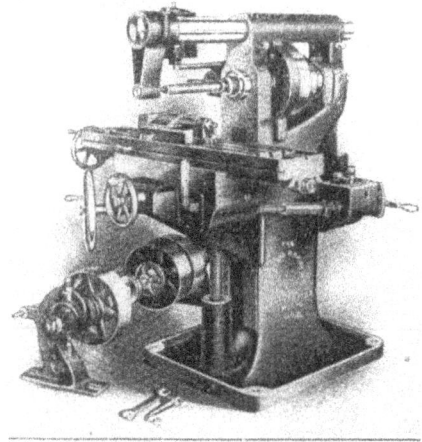

A PLAIN MILLING MACHINE

GRANT MACHINE TOOL WORKS,

Cleveland, OH

Founded July 30, 1898, by John J. Grant (1844-1934) after he had founded or worked for a number of companies. Products included a Grant-designed milling machine "for manufacturing operations such as are done on guns, sewing machines, typewriters, etc." The platen is 7" wide and 26" long with 16" of travel, *left.*

The company reorganized as the GRANT TOOL CO. in 1900 and moved to Franklin, PA.

GRAY MACHINE TOOL CO., Buffalo, NY

Maker, 1918-1920, of a line of thread milling machines, made in several sizes including the smallest *shown right.* The machine was designed to mill both internal and external threads.

Fig. 1. Smallest Size of the Gray Machine Tool Co.'s Thread Miller equipped with a Swivel Head

GRAYSON TOOL & MFG. CO., Indianapolis, IN

Formed in 1915 to make the TRIPLEX toolroom milling machine and shaper *shown left.* The machine featured a cylindrical column, telescoping for vertical adjustment, on which was mounted a heavy casting containing a vertical milling spindle, horizontal milling spindle and shaper. The desired function was manually rotated into the desired position.

HARDINGE BROTHERS, Chicago, IL
later Elmira, NY

Formed in July, 1890, by Franklin Hardinge and his brother Henry H. Hardinge. Early products were watchmakers' tools and small watchmakers' foot lathes. In 1902, the brothers bought the CATARACT line from the defunct CATARACT TOOL & OPTICAL CO. and began producing the line of bench lathes, milling attachments and milling machines. A new CATARACT bench milling machine (Fig.1) was introduced in 1906 and furnished with a small milling vise, dividing head and an adaptor for setting the dividing head to any angle up to 45 degrees. A larger bench milling machine (Fig.2), with vertical motion in a conventional knee, was introduced in 1913. Henry left the firm in 1895; Franklin was serving as chairman when the firm moved to Elmira, NY, in 1931.

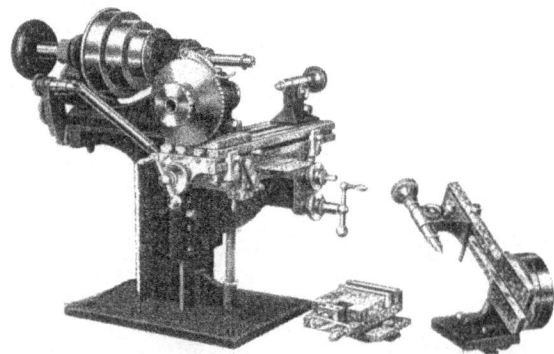

Fig.1

The Cataract Bench Milling Machine, Built by Hardinge Brothers, Chicago, Ill.

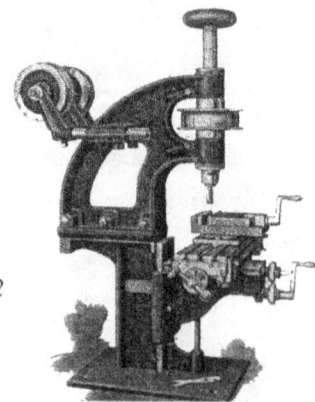

Fig.2

105

HARRINGTON MACHINE CO., Erie, PA

Builder, c1912-1914, of "under" milling machines. This type of milling machine was designed with the cutter placed under the work and was primarily used for trimming dies used in forging operations, *see right*.

HECKENDORN, H.C., Reading, PA

Formed in 1887 when the partnership of HECKENDORN & WILHELM broke up. Heckendorn produced milling machines, engine lathes and planers.

HECKENDORN & WILHELM, Reading, PA

A partnership of Henry C. Heckendorn and William H. Wilhelm, formed about 1880 to make light duty engine lathes and milling machines. The partnership was dissolved in 1887.

HENDERSON MACHINE TOOL CO., Philadelphia, PA

Formed in 1891 to make shapers and milling machines. Short lived, the company was reorganized as the PHILADELPHIA MACHINE TOOL CO. in 1894.

HENDEY MACHINE CO., Torrington, CT

Formed in 1870 as a partnership of Henry J. Hendey (1844-1906) and his brother Arthur Hendey (1846-1926). The firm was incorporated August 22, 1874. Arthur Hendey left in 1875. Henry J. Hendey continued to operate the firm and served as president until his death in 1906.

Shapers, planers and lathes were the primary products in the early years. Production of milling machines began about 1894 with the introduction of a continuous milling machine (Fig.1). Such machines were common 20 years later, but this was one of the first of its type. *(continued next three pages)*

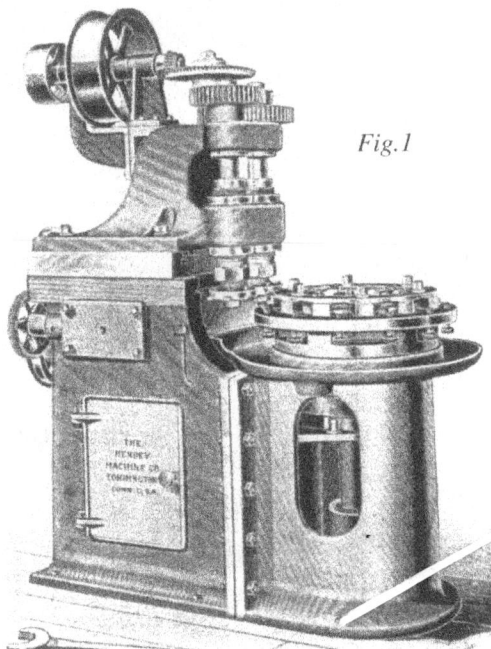

Fig.1

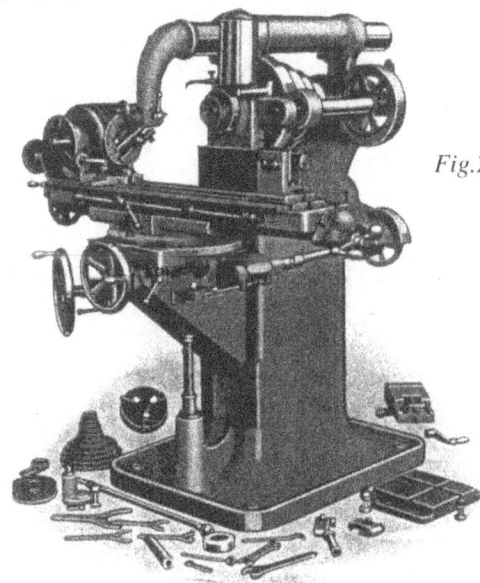

Fig.2

UNIVERSAL MILLING MACHINE—1900

Universal milling machines (Fig.2) were introduced in 1900 and joined in 1905 by No.2½ Lincoln milling machines (Fig.3). A peculiar oscillating milling machine was also introduced in 1905. Offered in No.2 (Fig.4) and No.4 (Fig.5) sizes, the machine was equipped with a spindle which oscillated back and forth rather than rotating. The design allowed "the cutter to be applied to the forming of slots, or the facing of interior parts on castings which would ordinarily be cut on shapers." The idea appears to have been short lived.

Fig.4

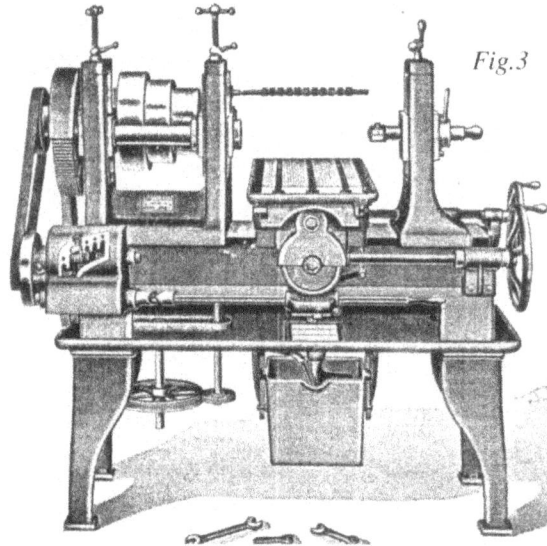

Fig.3

Fig.5

Fig.6

A new, all gear drive, universal milling machine (Figs.6-8) was also introduced in 1905. The speed and feed transmissions were of the type designed by Wendell P. Norton for the famous Hendey-Norton lathes. Cone drive, plain milling machines with the Norton gear box for feed only (Fig.9) were offered by 1907 and an improved No.4 universal milling machine (Fig.10) in 1910. A new dividing head (Fig.11) was also introduced in 1910.

Production of Lincoln milling machines continued and, by 1908 included the No.8 (Fig.12) weighing 4500 pounds and equipped with the Norton feed gear box. 1910 production included a double head model (Fig.13)

offered in six sizes and furnished with the Norton feed gear box. Production of cone drive milling machines (Fig.14) continued until 1920 or later.

Fig.7

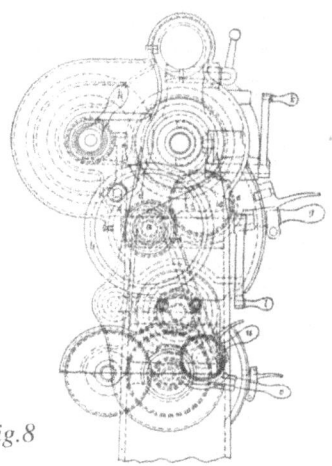

Fig.8

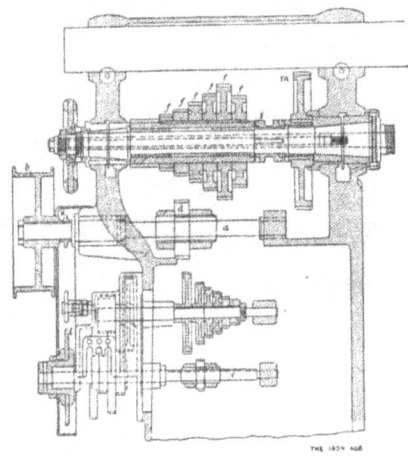

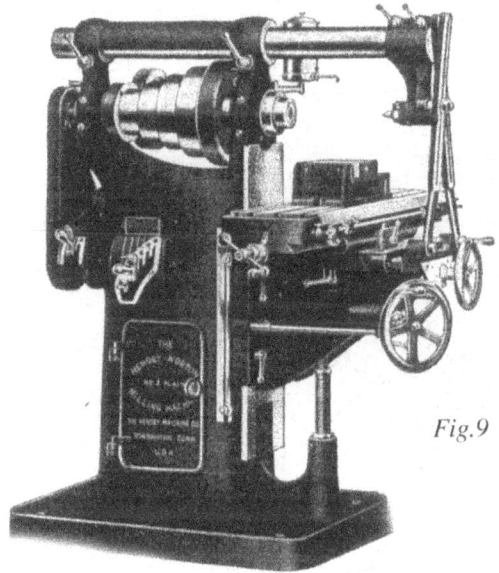

Fig.9

Feeds 28" x 8" x 19" All Automatic

Fig.10

Fig.11

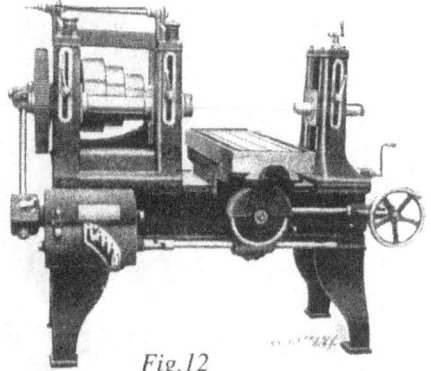

Fig.12

Fig.13

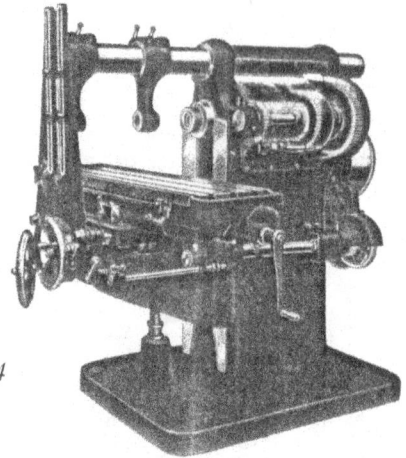

Fig.14

HESS MACHINE CO.,

Philadelphia, PA

Formed in 1902 by Henry Hess (1864-1922), the company was dissolved in 1904. Products included large, 36" wide, milling machines, *shown right,* as introduced in 1903. The machine set a new record by making a 20" wide cut in medium carbon steel, removing 3/8" stock at four inches per minute.

HETHINGTON & BERNER, Indianapolis, IN

Builder, in 1895, of column milling machines, used to face the ends of architectural and other structural iron beams or columns. Offered in two sizes, for columns 20' in length or 30' in length, the milling heads were 36" in diameter.

Columns Milling Machine.

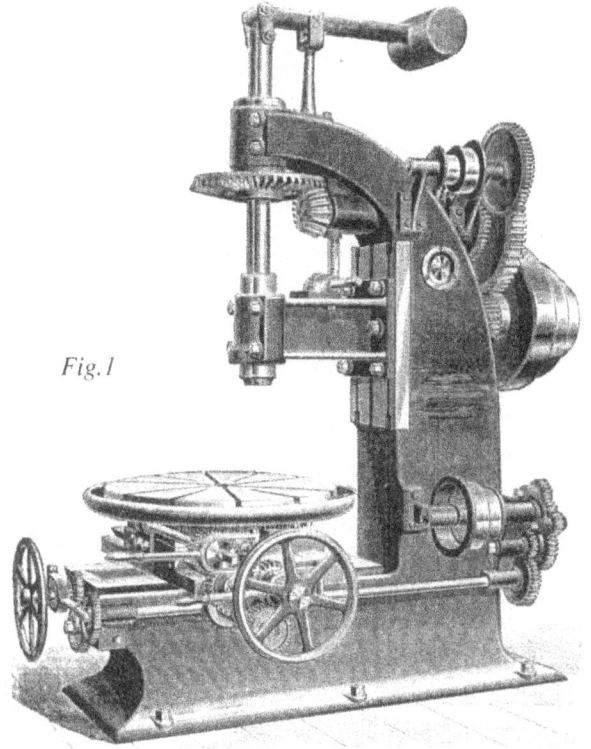

VERTICAL MILLING MACHINE.

Fig.1

HILLES & JONES, Wilmington, DE

A partnership of William S. Hilles (1823-1876) and Henry C. Jones (1838-1909), formed in 1870 when they bought out Robert H. Barr & Co. and reorganized as Hilles & Jones. The firm continued production of the Barr drilling machines. By 1883, vertical milling machines (Fig.1) were offered in four sizes from No.1 priced at $1,300 and weighing 6,560 pounds to No.4 priced at $1,900 and weighing 12,000 pounds. An improved version (Fig.2), equipped with a counterbalanced vertical slide and a hoist for loading workpieces was shown at the 1893 Columbian Exposition.

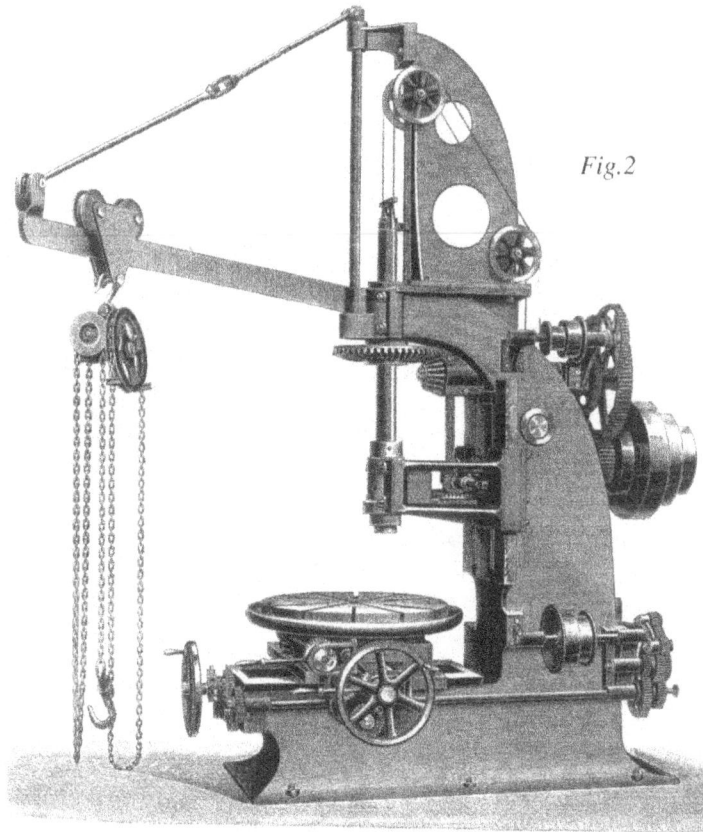

Fig.2

VERTICAL MILLING MACHINE.

HOE & CO., R., Brooklyn, NY

The largest American maker of printing presses, the firm also made milling machines and other machine tools during the Civil War.

IDEAL MACHINE CO., Hartford, CT

Maker, beginning in 1901, of the IDEAL index milling fixture, *shown at right*, designed as a cheaper alternative to the dividing head.

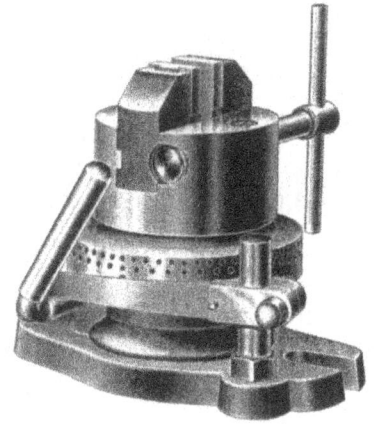

INGERSOLL MILLING MACHINE CO., Cleveland, OH, later Rockford, IL

Formed in 1890 by Winthrop Ingersoll (1865-1928) when he bought out his partner, Thomas Eynon, in EYNON & INGERSOLL. In April, 1891, Ingersoll moved the firm to Rockford, IL, where he served as president until his death in 1928.

Ingersoll continued production of slab milling machines, previously made by EYNON & INGERSOLL. A heavier version (Fig.1) was introduced by 1892 along with a combination boring and milling machine (Fig.2). By 1894, its ads (Fig.3) offered the slab millers in sizes from 14" square to 60" square. 1895 production included a 36" x 36" x 12' model with vertical spindle (Fig.4) and a 15" x 15" x 4' open side model (Fig.5). An improved version of the slab miller (Fig.6) was available by 1900.

(continued on next two pages)

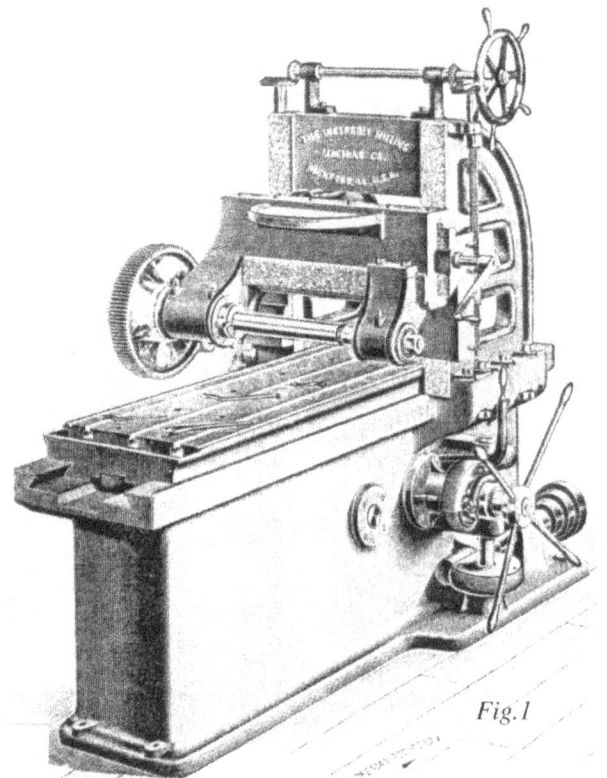

Fig.1

Fig.2

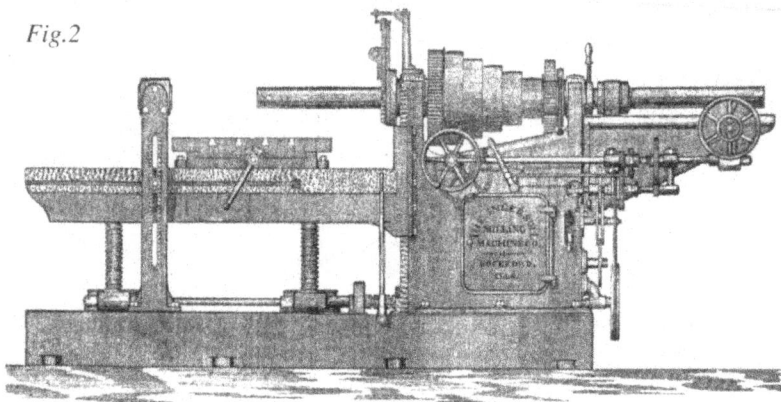

BORING AND MILLING MACHINE.

36-inch x 36-inch x 8-ft. Machine.

OTHER SIZES.

60 in. x 60 in. x 12 ft.
48 in. x 48 in. x 12 ft.
24 in. x 24 in. x 6 ft.
15 in. x 15 in. x 4 ft.

CROSS ADJUSTMENT.

BACK GEARED 24 : 1.

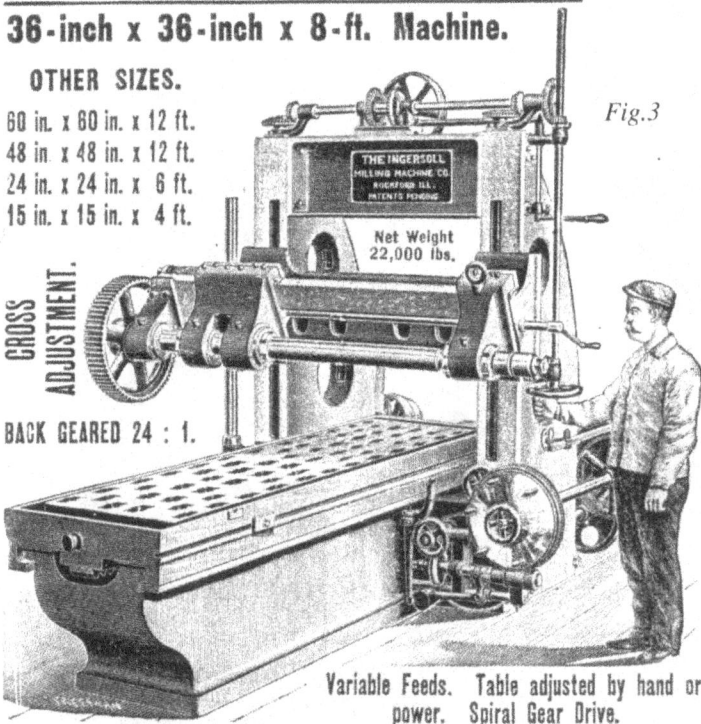

Net Weight 22,000 lbs.

Fig.3

Variable Feeds. Table adjusted by hand or power. Spiral Gear Drive.

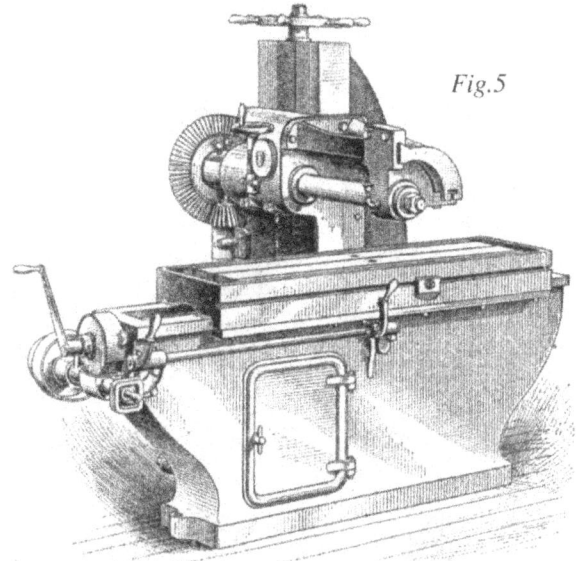

Fig.5

Open Side Slab Milling Machine.

HORIZONTAL OR VERTICAL SPINDLES, OR BOTH.

SIZES 15" X 15" X 4 FT. UP TO ANY REQUIRED WIDTH AND LENGTH.

Each spindle is independently driven. A cutter 2 inches diameter can be used in one at the same time that a cutter 18 inches diameter is used in the other. Automatic reversable feeds to table and vertical spindle.

Can furnish circular table and Profiling Attachment with this Type Machine.

Fig.4

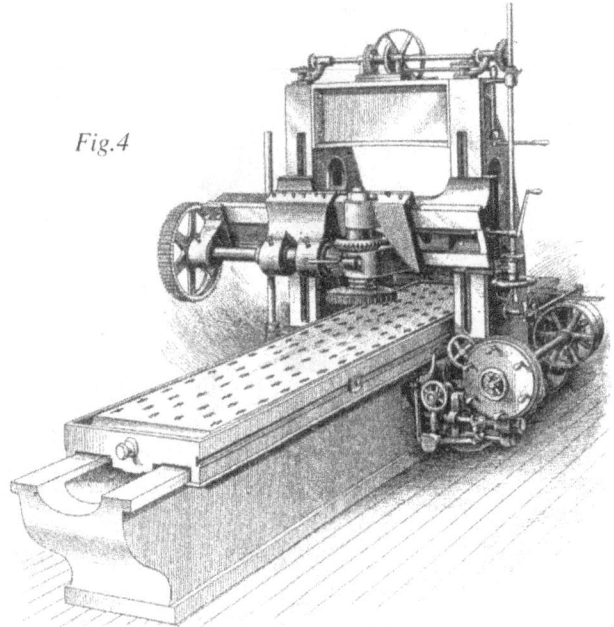

Slab Milling Machine.

Fig.6

Standard Combined Horizontal and Vertical Machine 4 in. x 32 in. x 8 ft.

Fig.8

Fig.7

Fig.10

Fig.9

Ingersoll continued to specialize in heavy milling machines, and by 1905 had built a 10', four head, planer type machine (Fig.7) A knee type, combined vertical and horizontal miller was introduced in 1909 and offered in No.3 (Fig.8) and No.5 (Fig.9) sizes, both with constant speed input. An improved model (Fig.10) was introduced in 1915. By the early 1930s, Ingersoll was building heavy milling machines weighing up to 275 tons.

JACKSON MACHINE TOOL CO., Jackson, MI

Formed in 1914 to take over production of the MELLING die-sinking machine previously made by the MELLING-NORTHUP CO. The machine, designed to cut semi-circular impressions (cherrying) in drop forge dies, was improved with the addition of power feeds and a stationary second spindle for regular milling operations.

DUPLEX *Typeless* Die Sinking Machine

And then not satisfied with that, we agree to demonstrate that dies such as crank and camshafts, knuckle joints, etc., can be machined in one setting, and all typing on semi-circular impressions entirely eliminated?

Will not the time-and-labor-saving possibilities cause you to investigate our proposition at once?

We are confident that it will, and that you will send for the circular which gives the full story of the Duplex Typeless Die Sinker—*today*.

Jackson Machine Tool Company
Jackson, Mich., U. S. A.

Cable:
"Die Sinker Jackson"

JAMES & FOOTE, Chicago, IL

A partnership of D.O. James and J.B. Foote (1865-1922) formed in 1893 to make gears. The firm was burned out in 1904 and thereafter split into the J.B. FOOTE GEAR & MACHINE CO. and D.O. James.

The only known machine production was a universal milling machine (*at right*) designed by J.B. Foote and introduced in late 1903. Production of the machine was continued by the J.B. FOOTE GEAR & MACHINE CO.

A NEW MILLING MACHINE

JOHNSON, ROBERT, Middletown, CT

An English gunmaker who came to the U.S. in 1814 where he opened a gun shop that made horse pistols and muskets for the U.S. military. Johnson is credited with making one of the earliest milling machines, in 1818. A drawing of the machine, *shown on next page*, was made by E.G. Parkhurst from memory in 1900. Parkhurst had been shown the machine in 1851 by Robert Johnson who described it as basically a rotary file used to rough out gun parts which were then hand filed to size. *(illustration on next page)*

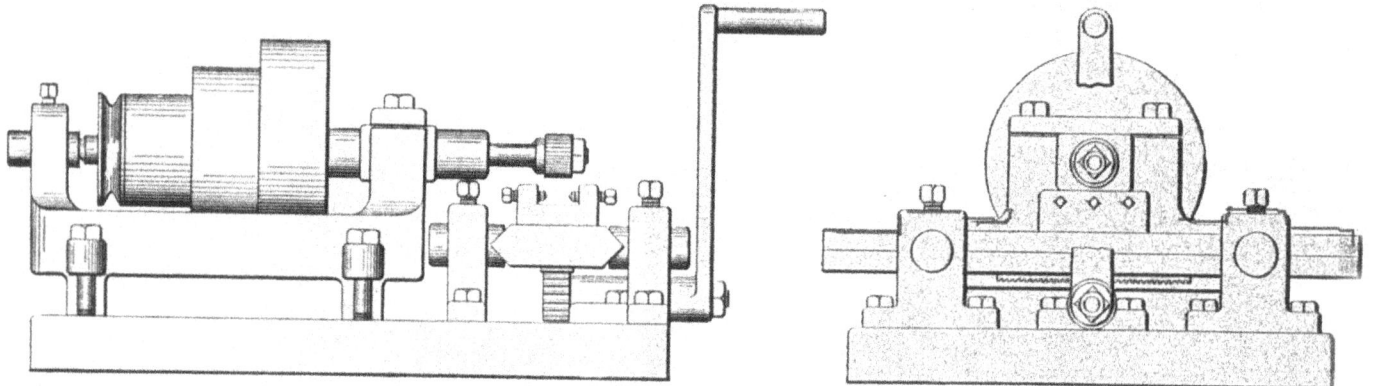

PRIMITIVE HAND MILLING MACHINE, MADE IN 1818.

American Machinist

JOHNSON TOOL CO., Providence, RI

Maker, beginning about 1920, of a cutter plate and die milling machine, furnished with a tilting table.

The Johnson Cutter Plate and Die Milling Machine

This machine has been designed for the rapid and accurate manufacture of punching dies and hubs.

Simplicity of design, convenience, extreme care in their manufacture and many other features make them the most efficient and reliable tools of their kind on the market today and place them in a class by themselves.

Machines are provided with the Tilting Table, a very valuable feature for this class of work, as it enables the operator to obtain a good view of the work while in progress. The machines have two spindles, one carried under the vise table, and the other above the table, which is carried in a swinging head fastened to the column of the machine, when not in use this spindle is swung out of position.

Machines are heavy, powerful and accurate and all material used in their construction is of the highest quality.

Johnson Tool Co., Inc.
201 Eddy St., Providence, R. I.

KEARNEY & TRECKER, Milwaukee, WI

KEARNEY & TRECKER CO., Milwaukee, WI

A partnership of Edward J. Kearney (1868-1934) and Theodore Trecker (1868-1935) formed in 1898. Both partners had worked at the KEMPSMITH MACHINE TOOL CO. where, so it is told, they made a pact that if Dewey won the battle of Manila Bay, they would leave and start their own company.

The firm incorporated in 1906 as the Kearney & Trecker Co. with Trecker as president until he retired in 1947 and Kearney as secretary and treasurer until his death in 1934.

First production appears to have been special drilling machines for bicycle makers. Milling machine production began in 1899 with the introduction of a No. 1 plain (Fig.1) with 28" table travel, 8" cross travel and 20" vertical travel. The line expanded in 1900 to include No. 2 universal millers (Fig.2) and No. 2 plain millers (Fig.3) offered with universal dividing heads (Fig.4), vertical spindle attachments (Fig.5), vertical spindle and rack cutting attachments with vise (Fig.6), and milling machine vises (Fig.7). Four models, all with automatic feed, were offered in 1903 (Fig.8), including No. 1 universal millers (Fig.9) and No. 2 plain millers (Fig.10).

In 1905, a new line of constant-speed drive, high power milling machines was introduced. This line was built in several sizes, including 2B plain (Fig.11) and 3B universal (Fig.12), and formed the heart of their milling machine production until 1912.

A universal cam cutting machine (Fig.13) was introduced in 1909 for cutting both face and periphery cams. Solid head vertical milling machines (Fig.14) were introduced in 1911. The trademark double overarm milling machine was introduced in 1912 in the 3B plain type (Fig.15), followed shortly by the 2B universal (Fig.16).1920 production included manufacturing milling machines Nos. 1A (Fig.17) with 22" x 8"x 19" travels, 1½A (1B (Fig.20) with 22" x 8" x 19" travels, 1½B (Fig.21) with 28" x 10" x 19" travels, 2B (Fig.22) with 28" x 10" x 19" travels; universal milling machine Nos. 1B (Fig.23) with 22" x 8" x 18" travels, 1½B (Fig.24) with 28" x 10"x 18" travels, 2B (Fig.25) with 28" x 10" x 18" travels, 3B (Fig.26) with 34" x 12" x 19" travels; and vertical milling machines Nos. 1½B (Fig.27) with 30" x 12" x 18" travels, 2½B (Fig.28) with 36" x 15" x 20" travel.

Attachments offered in 1920 included vertical spindle attachments (Fig.29), universal milling attachments (Fig.30), slotting attachments (Fig.31), universal centers (Fig.32), plain centers (Fig.33), spiral universal centers (Fig.34), rotary tables (Fig.35), rack cutting attachments (Fig.36), and rank indexing attachments (Fig.37).

(continued next six pages)

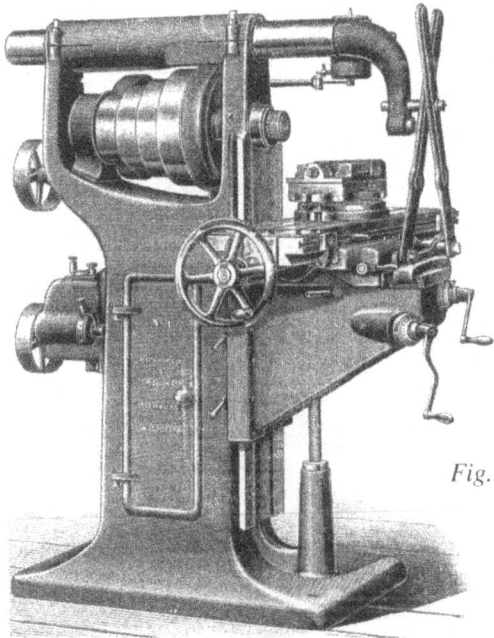

Fig.1

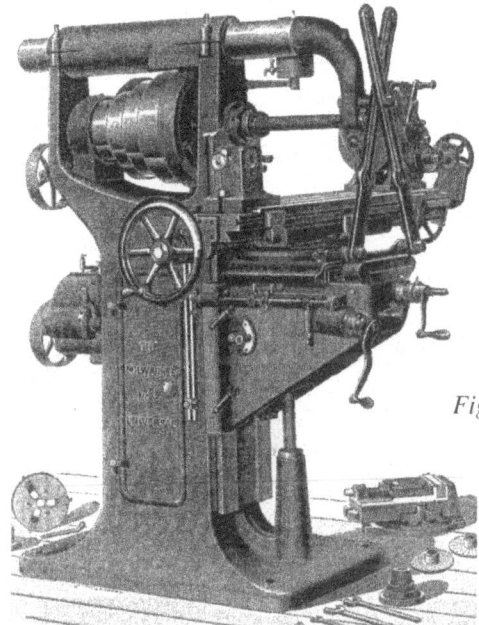

Fig.2

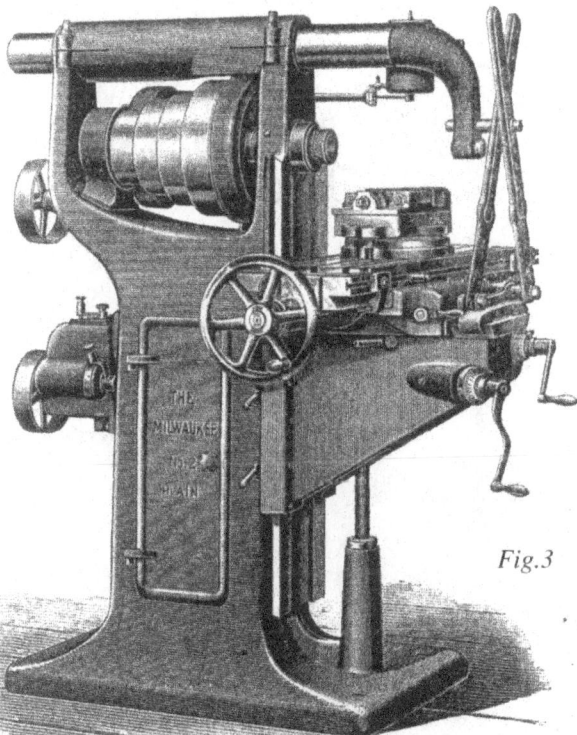

Fig.3

—Universal Dividing Head

Fig.4

—Vertical Spindle Attachment.

Fig.5

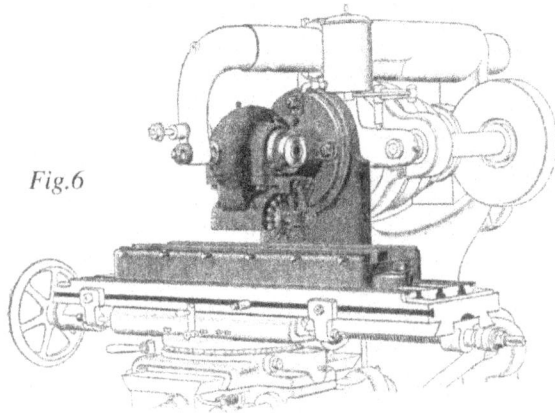

Fig.6

—*Vertical Spindle and Rack Cutting Attachment*

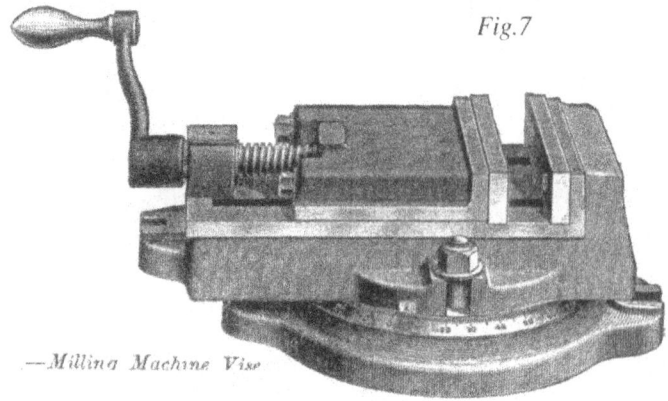

Fig.7

—*Milling Machine Vise*

SIZES AND DIMENSIONS

MACHINE	PLAIN		UNIVERSAL	
	No. 1	No. 2	No. 1	No. 2
Automatic table feed	24"	29"	24"	30"
Automatic cross feed	8"	10"	8"	10"
Automatic vertical feed	18"	20"	17"	18"
Working surface of table	8" x 37½"	10" x 44"	8" x 37½"	10" x 44"
Number of T slots in table	3	3	3	3
Width of T slots	11/16"	11/16"	11/16"	11/16"
Index centers swing			10"	12"
Index centers take in length			20"	25"
Number of steps on cone pulley	4	4	4	4
Largest diameter of cone pulley	10"	12¼"	10"	12¼"
Width of driving belt	2½"	3"	2½"	3"
Number of speeds direct, or direct and reverse	16-8	16-8	16-8	16-8
Size of swivel vise jaws	5¼" x 1⅜"	6¼" x 1½"	5¼" x 1⅜"	6¼" x 1½"
Vise opens without steel jaws	4¼"	5"	4¼"	5"
B. & S. taper hole in spindle	No. 10	No. 10	No. 10	No. 10
Number of feed changes	12	12	12	12
Variation in feed to one revolution of spindle	.006" to .130"	.006" to .130"	.006" to .130"	.006" to .130"
Width of feed belt	1¼"	2"	1¼"	2"
Size of countershaft pulleys	10" x 3"	12" x 3½"	10" x 3"	12" x 3½"
Speed of countershaft in same direction	140 & 170	140 & 170	140 & 170	140 & 170
Speed of countershaft in opposite directions	160	160	160	160
Net weight	1,800 lbs.	2,700 lbs.	2,000 lbs.	2,900 lbs.

Fig.8

Fig.9

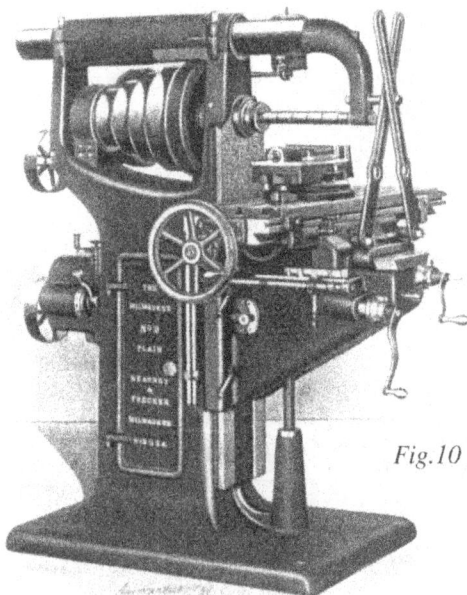

Fig.10

Fig.11

Fig.12

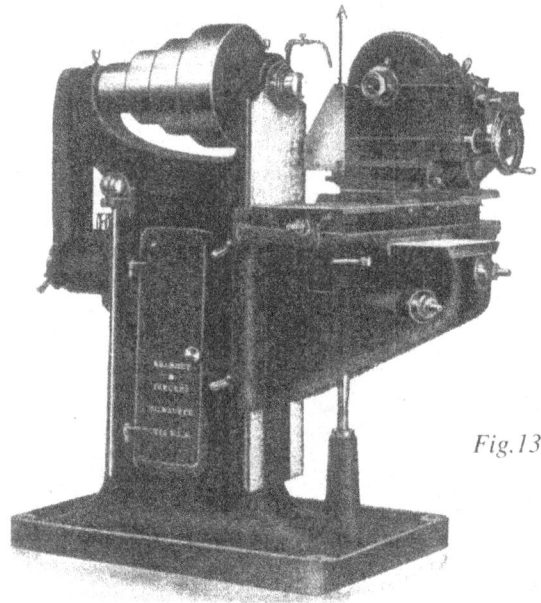

Fig.13

Fig.14

Fig.15

Fig.16

Fig.17

No. 1A MANUFACTURING MILLING MACHINE

Fig.18

No. 1½A SPECIAL MANUFACTURING MILLING MACHINE

Fig.19

No. 1A MANUFACTURING MILLING MACHINE

Fig.20

No. 1B PLAIN MILLING MACHINE

Fig.21

No. 1½B SPECIAL PLAIN MILLING MACHINE

Fig.22

No. 2B PLAIN MILLING MACHINE

Fig.23

No. 1B UNIVERSAL MILLING MACHINE

Fig.24

No. 1½B SPECIAL UNIVERSAL MILLING MACHINE

Fig.25

No. 2B UNIVERSAL MILLING MACHINE

Fig.26

No. 3B UNIVERSAL MILLING MACHINE

Fig.27

Fig.28

Fig.29

VERTICAL SPINDLE ATTACHMENT

Fig.30

UNIVERSAL MILLING ATTACHMENT

Fig.31

SLOTTING ATTACHMENT

Fig.32

Fig.33

PLAIN CENTERS

Fig.34

Fig.35

ROTARY TABLE

Fig.36

RACK CUTTING ATTACHMENT AT WORK

Fig.37

Kearney & Trecker

RACK INDEXING ATTACHMENT

KEATING & MARQUARDT, New Haven, CT

A partnership of J.J. Keating and F.C. Marquardt, formed in 1907 to make a cam milling machine. The very complex machine could be configured to cut face cams (Fig.1) or periphery cams (Fig.2) as required.

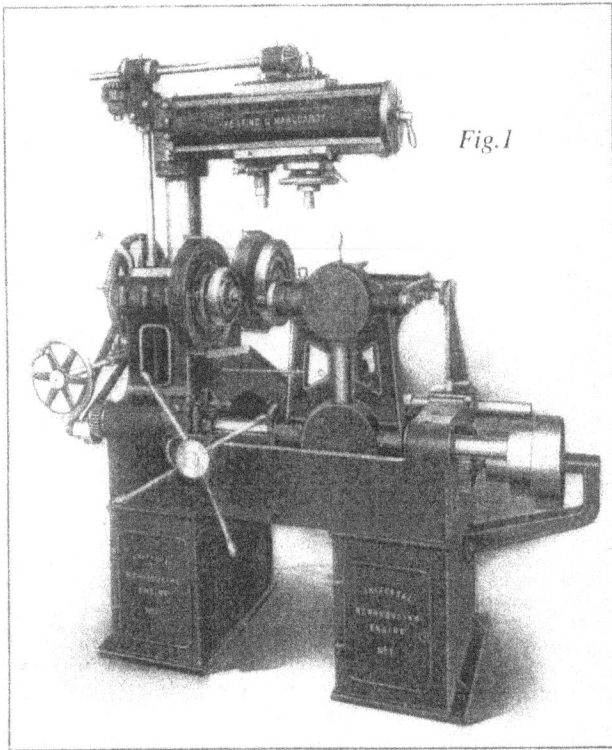

Fig.1

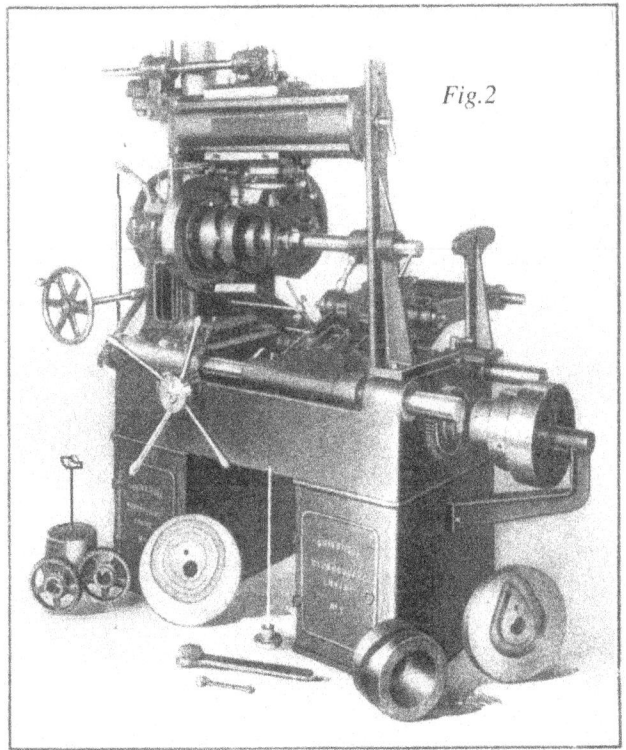

Fig.2

KELLER MECHANICAL ENGRAVING CO., Brooklyn, NY

Formed in 1896 by Sidney A. Keller (1872-1930) and his cousin Joseph F. Keller. Early production was limited to engraving dies for the silverware and jewelry trade. In 1909, Joseph F. Keller developed a die engraving machine which utilized a mechanical tracer. By 1914, the engraving machines became die milling machines and were offered in a range of sizes for use on dies up to large drop forging sizes *(see next page)*. The firm reorganized as the Keller Mechanical Engineering Co. in 1922 and sold out to the Pratt & Whitney Co. in 1931.

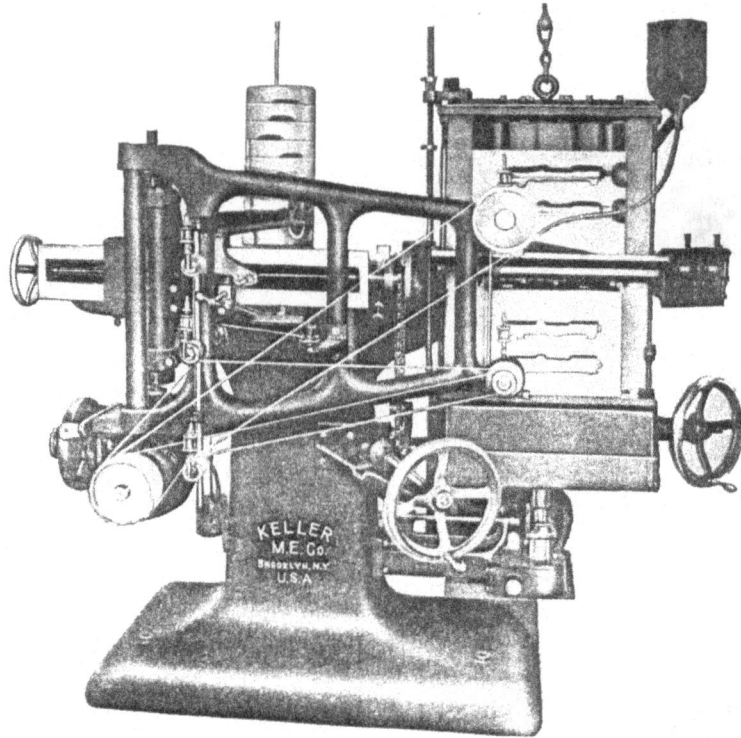

Keller

KELLY, PATRICK J., Elizabeth, NJ

Builder, in 1892, of a milling machine patented October 27, 1891, and designed for straddle milling of workpieces such as steam radiator sections (Fig.1). Note the unusually high gear reduction to the feed mechanism (Figs.2-3).

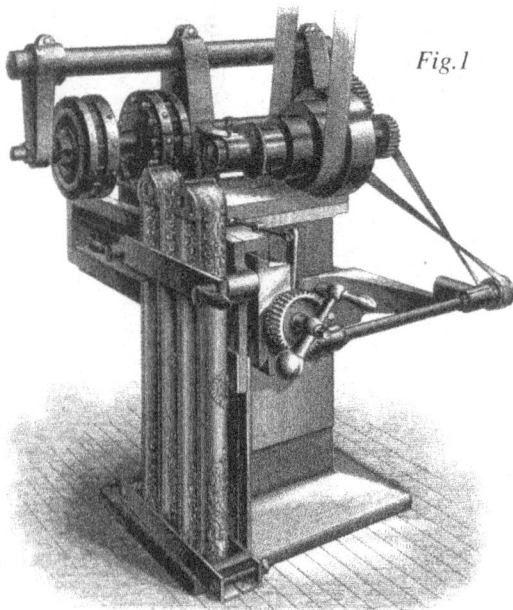

Fig.1

THE KELLY MILLING MACHINE.

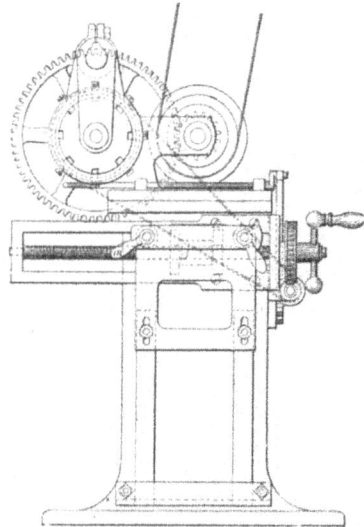

Fig. 2.—End Elevation.

Fig.2

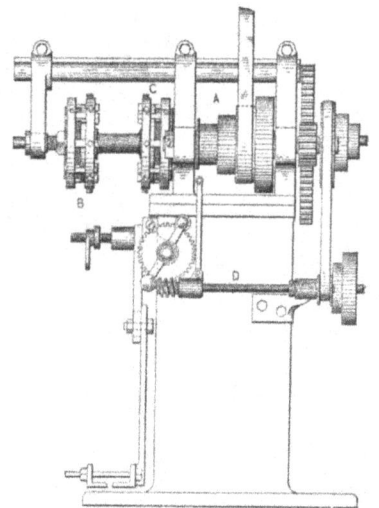

Fig. 3.—Side Elevation.

Fig.3

KEMPSMITH MACHINE TOOL CO., Milwaukee, WI
later KEMPSMITH MFG. CO., Milwaukee, WI

Formed by Frank Kempsmith (1848-1904) in 1889 as a reorganization of the MILWAUKEE MACHINE TOOL CO. Kempsmith, who had been a BROWN & SHARPE MFG. CO. apprentice, was also co-founder of the Springfield Machine Tool Co. in 1887, but left in 1888. Because of poor health, Kempsmith sold his company to Howard L. Browning and Franz Wollaeger (1847-1935) in 1901. The new owners reorganized as the Kempsmith Mfg. Co. with Browning as president and Wollaeger as secretary and treasurer. Early production was sold through LODGE & DAVIS which marketed the machines under its name (Fig.1). By 1890, Kempsmith was building machines under his own name, including a universal milling machine (Fig.2) with 24" table travel, 6" cross travel and 15" vertical travel. Furnished with a dividing head, the machine weighed 2,200 pounds complete.

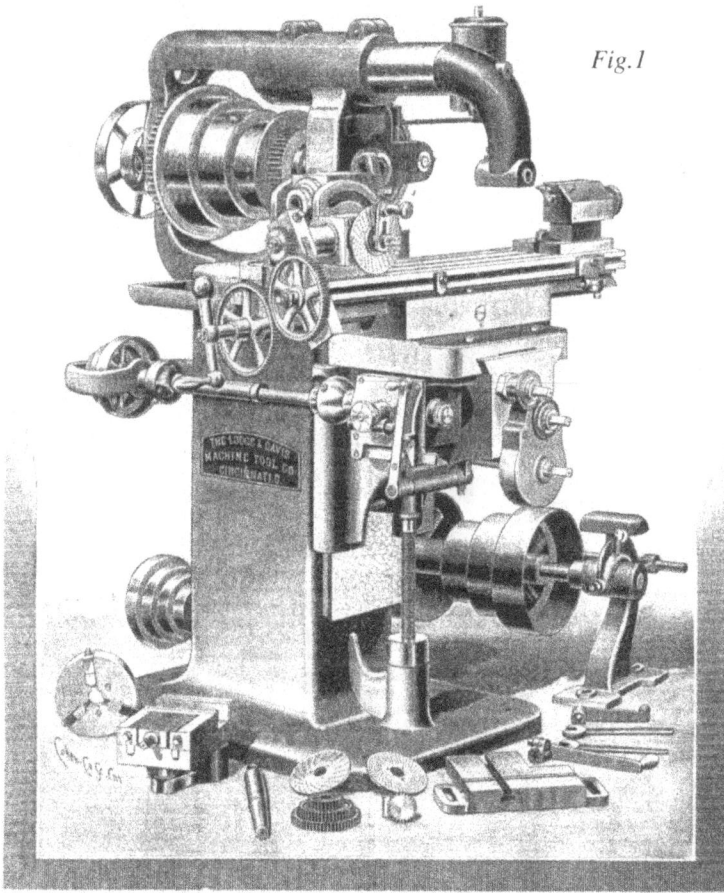

Fig.1

THE KEMPSMITH IMPROVED UNIVERSAL MILLING MACHINE.

Fig.2

NEW UNIVERSAL MILLING MACHINE.

A new line of milling machines was introduced in 1892 and included No. 2 plain milling machines (Fig.3) with 18" table travel, $3\frac{3}{4}$" cross travel and $14\frac{1}{2}$" vertical travel. 1894 production included a new No. 0 universal milling machine (Fig.4) with 18" table travel, $4\frac{1}{2}$" cross travel and $14\frac{1}{2}$" vertical travel, furnished with a new design dividing head (Fig.5); and a new No. 5 plain (Fig.6) with 44" table travel, 10" cross travel and 21" vertical travel, weighing 3,600 pounds. *(continued next three pages)*

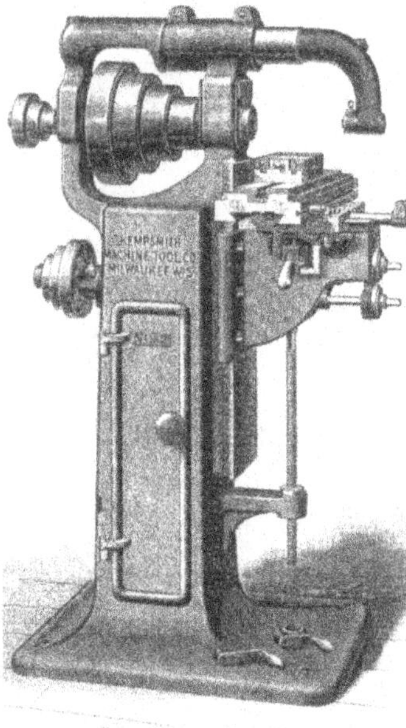

Fig.3

PLAIN MILLING MACHINE

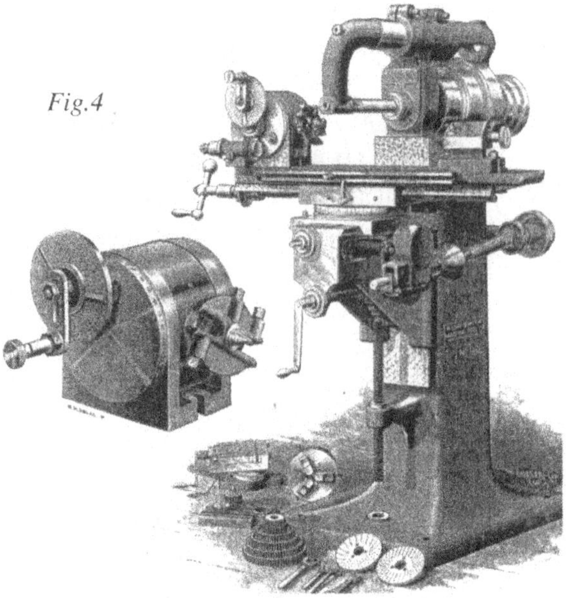

Fig.4

Fig.5

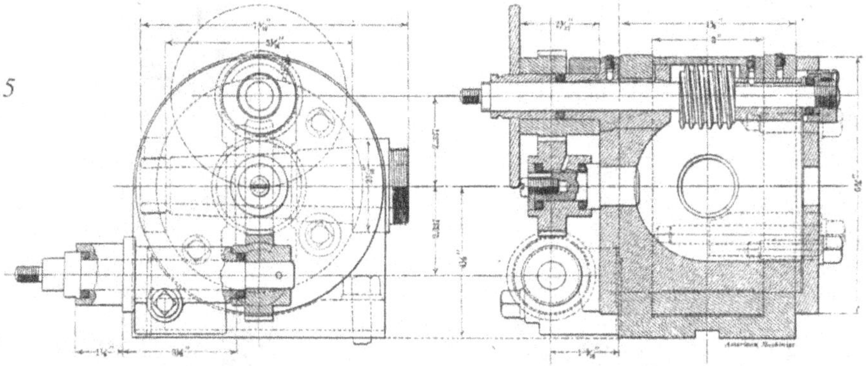

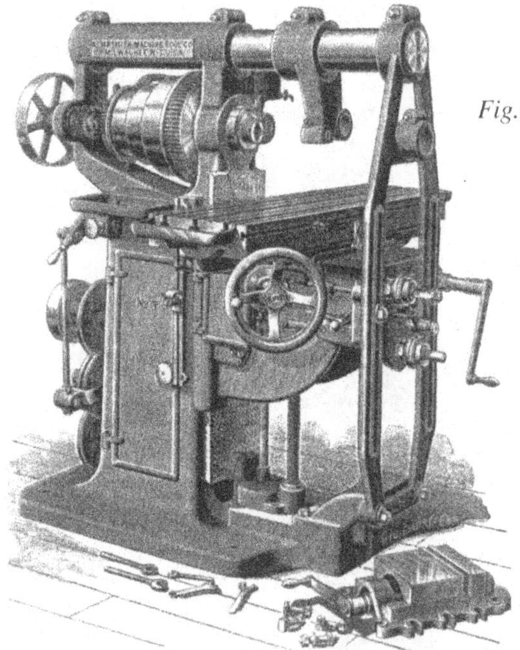

Fig.6

A NEW MILLING MACHINE

Fig.7

UNIVERSAL MILLING MACHINE

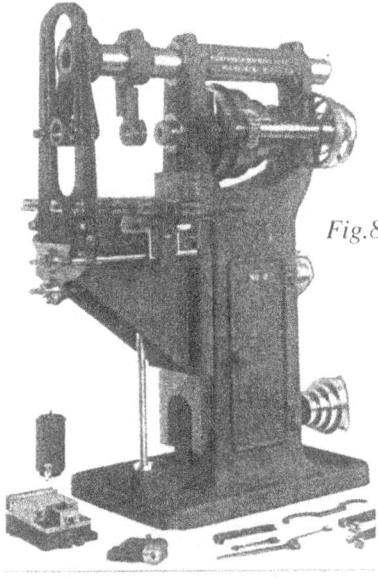

Fig.8

PLAIN BACK-GEARED MILLING MACHINE.

Fig.9

IMPROVED MILLING MACHINE

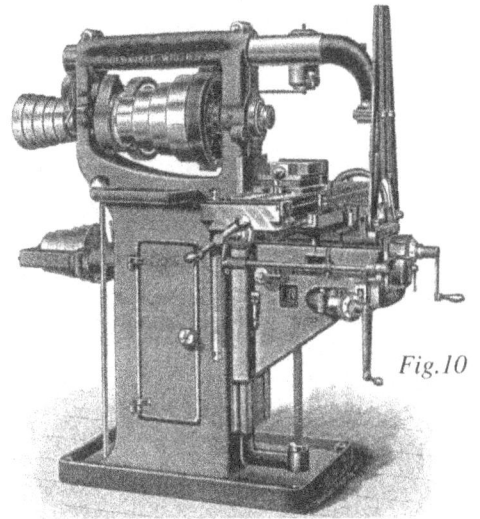

Fig.10

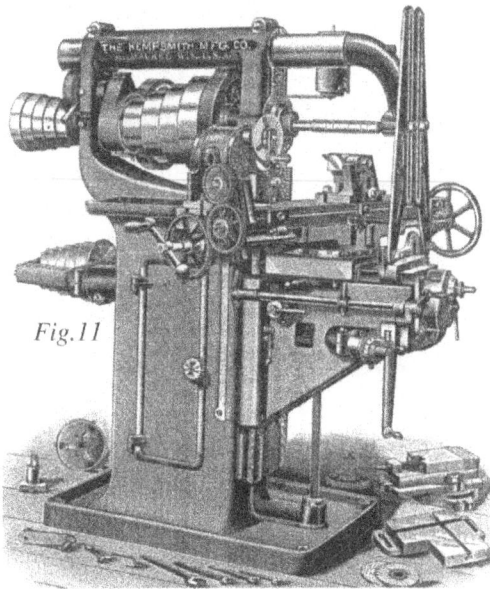

Fig.11

Fig.12

THE KEMPSMITH NEW UNIVERSAL MILLER

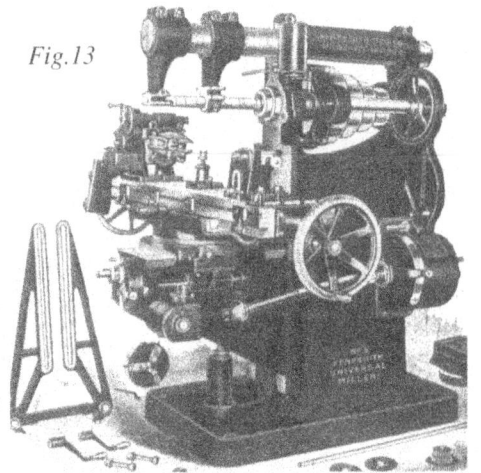

Fig.13

Fig.14

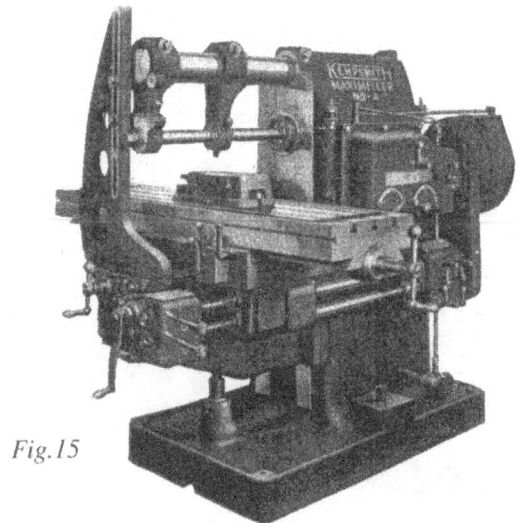

Fig.15

A No. 8 universal miller (Fig.7) with 24" table travel, 7" cross travel and 19¾" vertical travel was introduced in 1896, along with the No. 9 plain, back-geared milling machine (Fig.8). 1898 production included an improved Lincoln type miller (Fig.9) with a 12' x 38" table. 1901 production included the No. 1½ plain milling machine (Fig.10) and No. 3 universal (Fig.11), two of a total of thirteen styles made by Kempsmith at that time. The universal milling machine was redesigned in 1903, fitted with back gears and a double friction countershaft and designated the No. 22 (Fig.12). By 1905, the machine was offered in No. 3 (Fig.13) size.

Another redesign of the Lincoln miller (Fig.14), now equipped with a motor drive, was introduced in 1913. The MAXIMILLER was introduced in 1918 (Fig.15) and became the heart of the Kempsmith milling machine line into the 1930s.

KENT-OWENS MACHINE CO., Toledo, OH

Formed in 1912 to make bottle making machinery developed by Michael J. Owens (1859-1923). Milling machine production began in 1915 with the introduction of a high-speed hand miller, *shown at right*. The machine weighed 1,080 pounds and had 17" table travel and 5" vertical travel.

KEYSTONE MACHINE TOOL WORKS, Philadelphia, PA

Formed in 1901 as a reorganization of F.H. Gleim & Co. The firm continued production of Gleim cold saw machines and horizontal boring machines, adding, by 1903, heavy milling machines as *shown below*.

HIGH-SPEED HAND MILLER

Keystone Special
MILLING MACHINE

Designed for Heavy and Exacting Work

Box Pattern Housings and Bed.
Main Spindle Bearings, 5½ x 22 inches
16 Automatic Feed Changes.
Power Return 10 ft. per minute.

Belt or Motor Driven.

We also build

Cold Saw Cutting-
Off Machines

Horizontal
Boring Machines

Horizontal Floor
Boring Machines

Slab Milling Machines

Automatic
Saw Sharpeners

Write
for Catalog

KEYSTONE MACHINE TOOL WORKS
Ridge Ave. and Hamilton St.   PHILADELPHIA, PA.

KNIGHT MACHINERY CO., W.B., St. Louis, MO

Founded in 1891 by William B. Knight (1866-1950) to make bicycles. Knight incorporated as the W.B. Knight Machine Co. in 1905 and began production of combined drilling and milling machines in 1906.

The first combination drilling and milling machine (Fig.1), later designated the No. 1, was made with a table travel of 12" and cross travel of 6". Milling vises and rotary tables were offered with the machine. In 1909, a heavier, No. 2 version (Fig.2) with 20" table travel and 8" cross travel was introduced, along with an improved model of the No. 1 (Fig.3). By 1911, a No. 1½ model (Fig.4) with 16½" table travel and 8" cross travel was available.

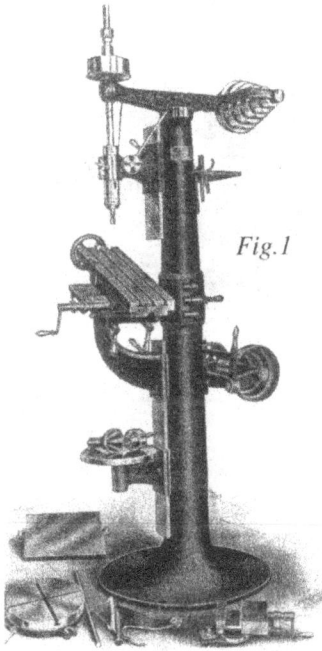

Fig.1

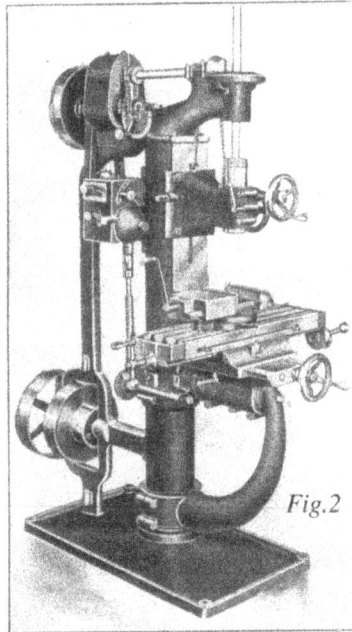

Fig.2

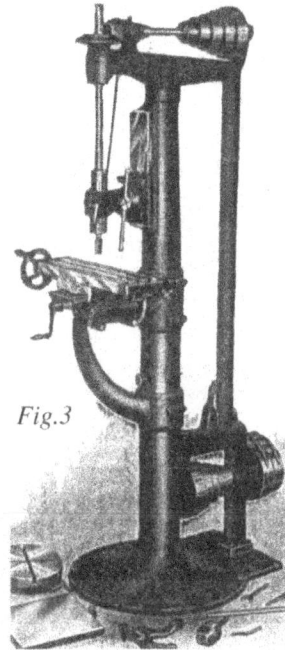

Fig.3

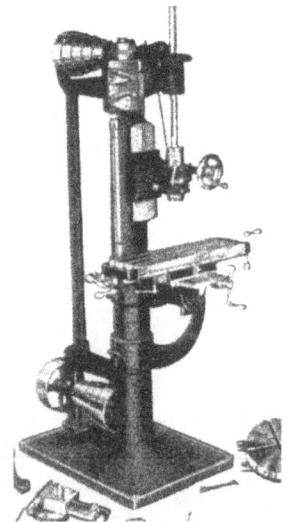

Fig.4

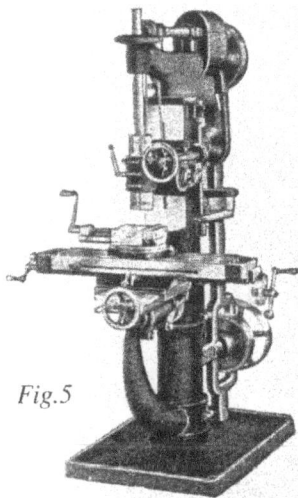

Fig.5

Fig.6

1920 production included a No. 3 model (Fig.5) with 24" table travel and 10½" cross travel; and a duplex milling machine (Fig.6) of much heavier construction than the vertical machines. The firm continued to operate into the 1960s.

KONIGSLOW, OTTO, Cleveland, OH

Established in 1875, Konigslow was the maker, in 1883, of light machinery including the small milling machine *shown at right*. The design probably dates to c1875.

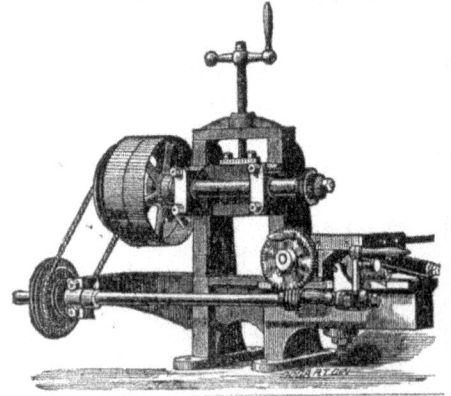

KUNZ MACHINERY CO., J.L., Milwaukee, WI

Builder, 1915-1917, of a small milling machine (Fig.1). It was offered with a rather complex gear cutting attachment (Fig.2) and equally complex keyseating attachment (Fig.3). *(shown below)*

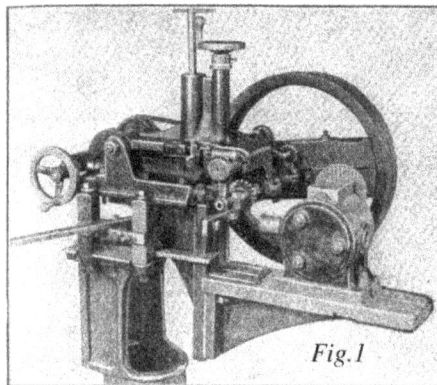

Fig.1

Fig.2

Fig.3

A Manufacturing Milling Machine

LAMB KNITTING MACHINE CO.,

Chicopee Falls, MA

One of the major manufacturers of knitting machines, the firm also offered bed type milling machines "a number of which have been built for its own use." In 1910 it began to build them for the market, to take work 48" long by 13" wide by 13" high. Three spindle speeds and three power feeds were standard.

LAMSON, GOODNOW & YALE, Windsor, VT

later **LAMSON & CO., E.G.,** Windsor, VT

Formed in 1859 when Ebenezer G. Lamson (1814-1892), Abel F. Goodnow and B. Buchcanan Yale bought the plant and equipment of the defunct ROBBINS & LAWRENCE CO. Henry D. Stone (1815-1899) who had worked at Robbins & Lawrence since 1847, was appointed superintendent continuing through the later reorganizations. In 1862, the firm reported employment of "100 men making gun machinery." A variety of other machinery was also offered. Lamson bought out his partners and reorganized as E.G. LAMSON & CO. in 1864. In 1865 he again reorganized, this time as the Windsor Mfg. Co. All the firms made engine lathes, turret lathes, and index mmilling machines based on Frederick W. Howe and Henry D. Stone designs originally done for ROBBINS & LAWRENCE.

LATHE & MORSE, Worcester, MA

later **LATHE & MORSE TOOL CO.,** Worcester, MA

A partnership of Martin Lathe and Edwin Morse formed in 1864 as a successor to SHEPARD, LATHE & MORSE. In 1871, the firm reorganized as a stock company, the Lathe & Morse Tool Co. Employment in 1882 was 40 to 50 men. The company was defunct in March, 1891, when the assets were sold to William F. Draper who then formed the DRAPER MACHINE TOOL CO. Products included a variety of machine tools, primarily planers and lathes. Milling machines, made on designs carried over from Shepard, Lathe & Morse, were offered as late as 1872 when they were priced at $350.

LEA, Edward S., Trenton, NJ

Builder, beginning in 1911, of milling machines designed to be quickly changed from a vertical to a horizontal. They were equipped with a head that rotated between vertical and horizontal and an unusual arbor support which allowed the vertical/horizontal conversion.

LEBLOND, R.K., Cincinnati, OH

later **LEBLOND MACHINE TOOL CO., R.K.,** Cincinnati, OH

Founded in 1887 by Richard K. LeBlond (1864-1953) to make gauges and small tools. In 1891, he began making 16" engine lathes on contract for LODGE & DAVIS MACHINE TOOL CO. In 1896, the Lodge & Davis contract ended and LeBlond began making 12", 14" and 16" engine lathes under his own mame. The firm incorporated as the R.K. LEBLOND MACHINE TOOL CO. in 1898 with a capital of $50,000 and engine lathes as the primary product.

Milling machine production began in 1900 with the introduction of a No. 2 plain (Fig.1). The milling machine line was redesigned and expanded in 1903 and now included No. 0, No. 1, No. 1 1/2, No. 2, and No. 3 plain (Fig.2) types, and No. 3 universal type (Fig.3). An improved No. 3 universal (Fig.4) was offered in 1907.

A linc of constant speed milling machines was introduced in 1908 and included motor drive (Fig.5) and single pulley drive (Fig.6) types. Heavy duty cone type millers (Fig.7) were available in 1912 with a quick power traverse feature (Fig.8) added in 1913. Heavy duty single pulley millers were introduced in 1913 in both plain (Fig.9) and universal (Fig.10) types. Milling machine production appears to have ceased during World War I, probably due to the great demand for engine lathes. *(illustrations continued on next two pages)*

Fig.1

No. 3 Plain Machine.   Milling 10½ x 20 x 36.

Fig.2

Fig.3

Fig.4

Fig.5

Fig.6

Fig.7

Fig.8

Fig.9

HEAVY-DUTY PLAIN MILLER WITH SINGLE
PULLEY DRIVE AND SELF-ALIGNING ARBOR SUPPORTS

Fig.10

HEAVY-DUTY UNIVERSAL MILLER WITH SINGLE-
PULLEY DRIVE AND SELF-ALIGNING ARBOR SUPPORTS

LINCOLN & CO., Hartford, CT   See PHOENIX IRON WORKS

LIPE, C.E., Syracuse, NY

Founded in 1882 by Charles E. Lipe (1851-1895) to make milling machines of the heaviest pattern offered to that time.

Machine production began with a 2800 pound, universal milling machine (Fig.1) introduced in 1882, complete with index head and center post (Fig.2). An enlarged and improved version (Fig.3), called Lipe's Heavy Universal Milling Machine, was introduced in 1884 as was a plain version (Figs.4-5). A set of "Ever-Day" centers (Fig.6) were introduced in 1885 as an alternative to the more expensive index heads. The universal machines and index heads were patented by Lipe February 5, 1884.

Production of the above machines continued until 1885 when Lipe sold the line and patent to the BROWN & SHARPE MFG. CO., which continued production for a time. Lipe's last known milling machine was a heavy vertical machine (Fig.7) introduced in 1892. *(illustrations continued on next two pages)*

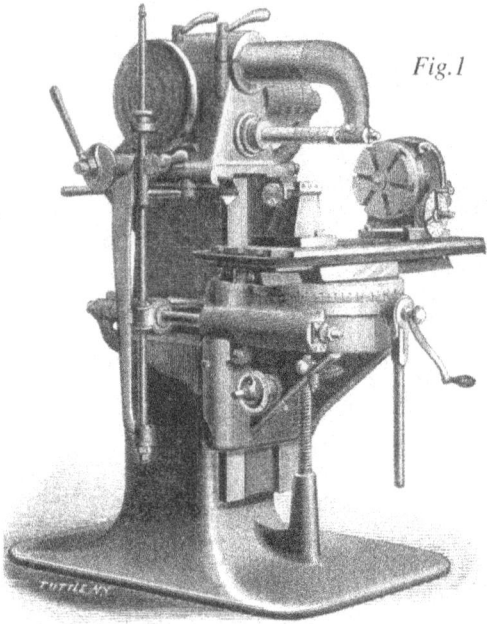

Fig.1

New Universal Milling Machine.

Fig.2

Index Head Centers and Center Rest on Plate.

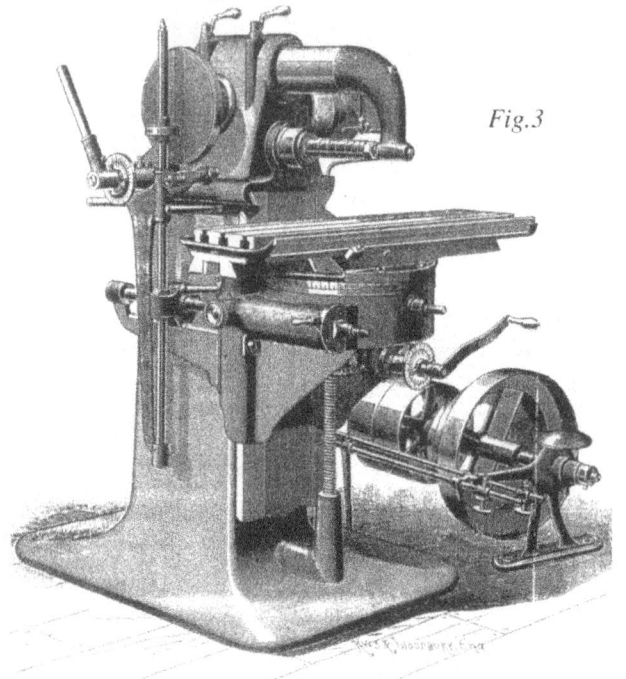

Fig.3

Lipe's Heavy Universal Milling Machine.

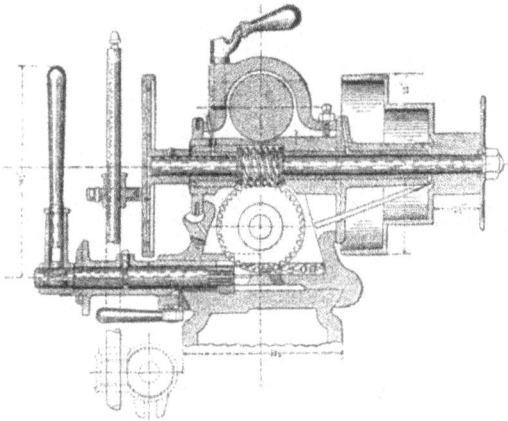

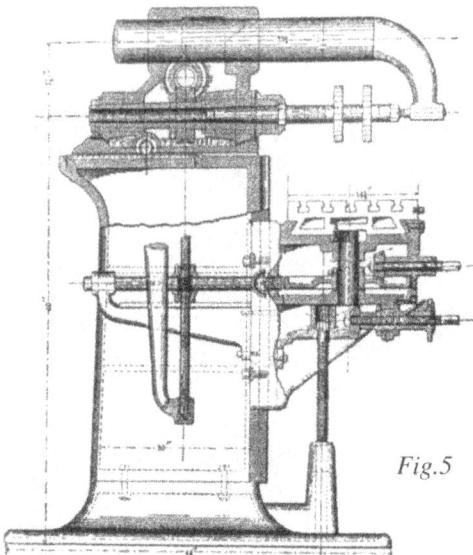

Fig.5

Fig.4

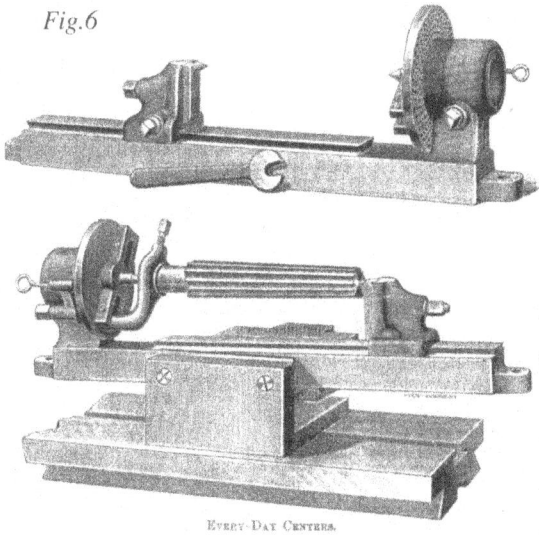

Fig.6

EVERY-DAY CENTERS.

Fig.7

SPECIAL VERTICAL MILLING MACHINE.

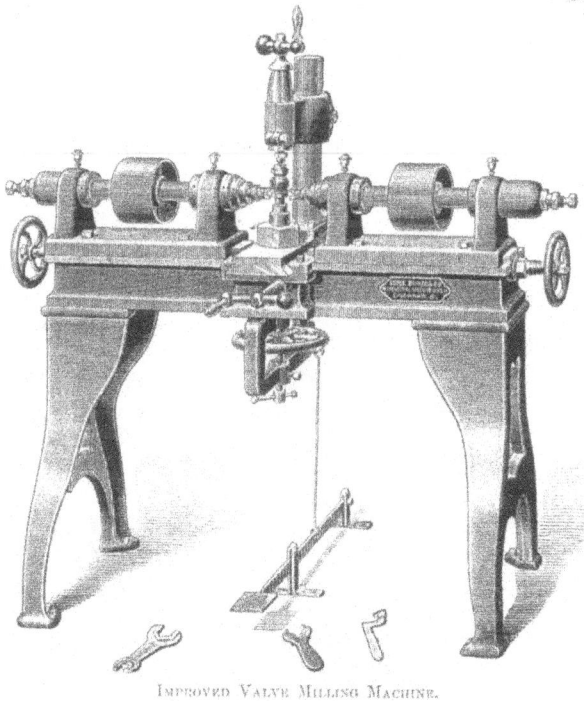

IMPROVED VALVE MILLING MACHINE.

LODGE, BARKER & CO., Cincinnati, OH

A partnership of William Lodge (1848-1917) and William Barker (1853-1915) formed January 1, 1880. Both Lodge and Barker had previously worked for the John Steptoe Co. The firm was reorganized as LODGE, DAVIS & CO. when Charles Davis replaced Barker in January, 1886. Engine lathes and brass finishers' machinery were the primary products, but the firm also offered valve milling machines as *shown at left*.

LODGE, DAVIS & CO., Cincinnati, OH
later LODGE & DAVIS MACHINE TOOL CO., Cincinnati, OH

A partnership of William Lodge (1848-1817) and Charles Davis (1853-1903) formed in January, 1886, as a reorganization of Lodge, Barker & Co. In 1889, the firm became a stock company, the Lodge & Davis Machine Tool Co. Lodge left the firm in March, 1892, to found the Ohio Machine Tool Works. Davis, who became president when Lodge departed, reorganized as the DAVIS & EGAN MACHINE TOOL CO. September 1, 1896, when Thomas P. Egan bought an interest. Henry Dreses (1854-1930), who had joined the firm in 1883, was chief designer by 1892, serving until 1896 when he left to form DRESES, MUELLER & CO. *(continued)*

The new firm continued production of the valve milling machine made by Lodge & Barker. A line of knee type milling machines was introduced in 1889 and offered in plain and back-geared (Fig.1) styles. Slab milling machines (Fig.2) were introduced in 1892 along with a KEMPSMITH universal milling machine (Fig.3) made by the KEMPSMITH MACHINE TOOL CO. but sold under the Lodge & Davis name.

A complete new line of milling machines, designed by Henry Dreses, was introduced in 1893. The new line included universal milling machines (Fig.4), heavy plain milling machines (Fig.5), and plain milling machines (Fig.6). An improved version of the universal milling machine (Fig.7) was introduced in 1894. *(continued)*

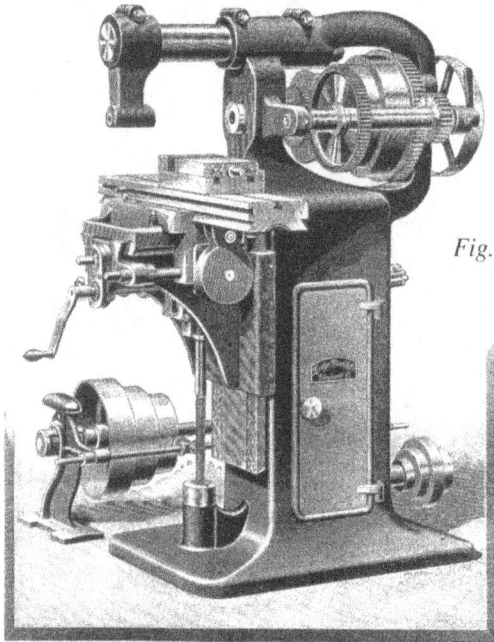

Fig.1

Fig.2

Slab Milling Machine.

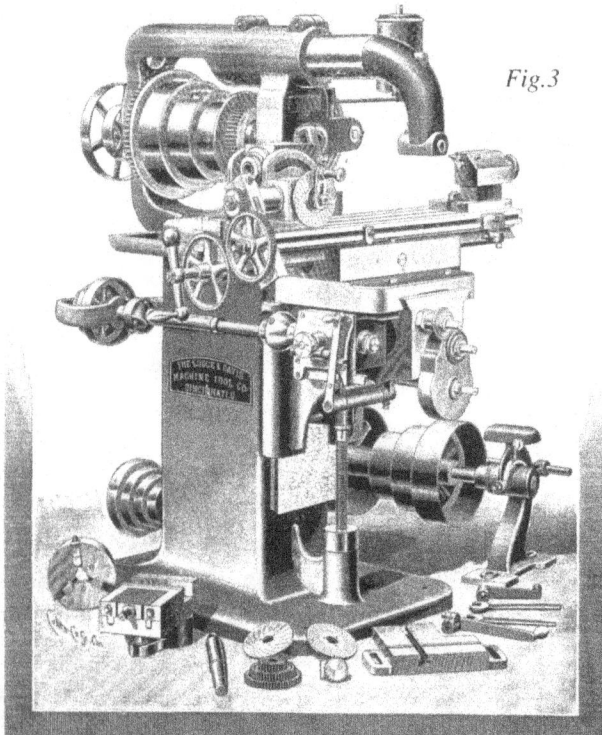

Improved Milling Machine.

Fig.3

Fig.4

The Kempsmith Improved Universal Milling Machine.

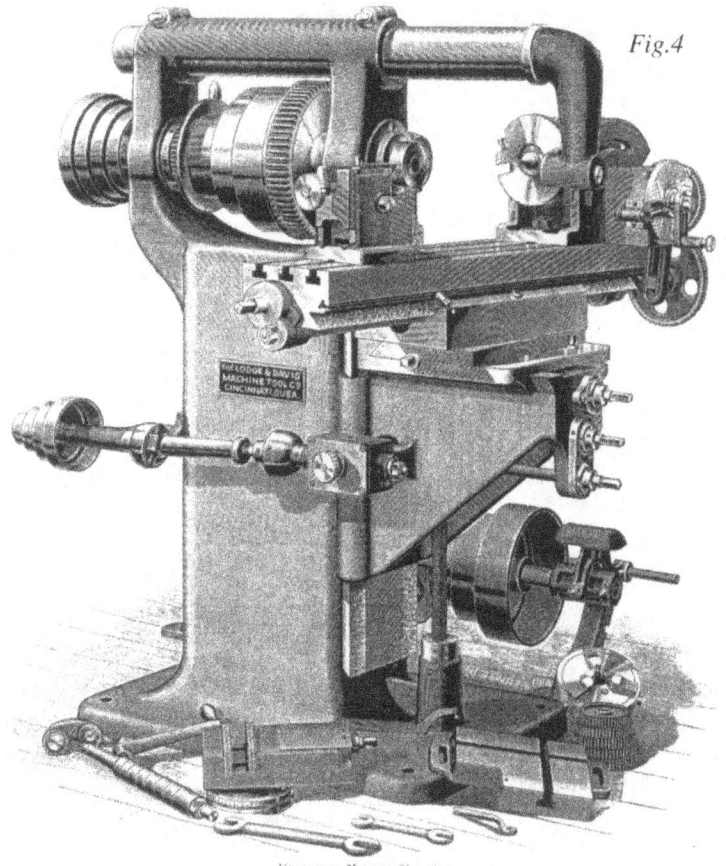

Universal Milling Machine.

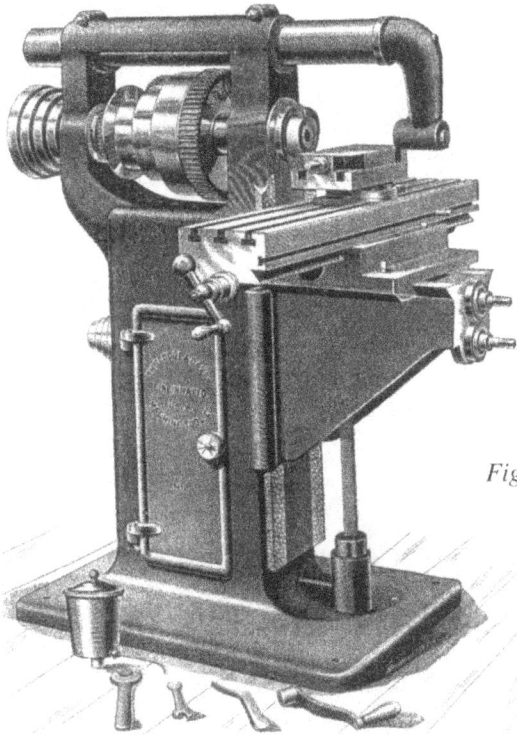

Fig.5

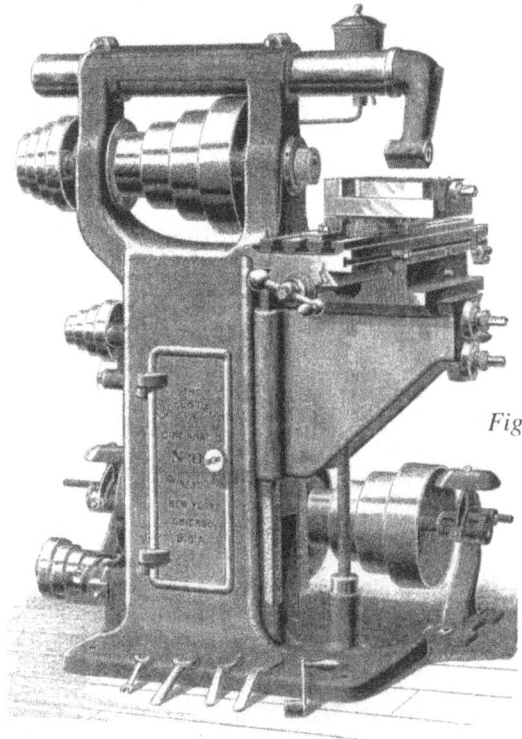

Fig.6

Fig.7

MANHATTAN FIREARMS CO., Newark, NJ

A large maker of pistols and revolvers formed in 1855. The firm began making milling machines for sale by 1863, continuing through the Civil War. The Andrew R. Arnold milling machine patent of August 30, 1864, was assigned to Manhattan as was the L.B. Miller patent of October 1, 1867, which covered a machine for milling twist drills.

MANSFIELD, WALLACE H., New Haven, CT

Builder, beginning in 1895, of 10" footpower engine lathes "adapted for milling". The milling feature consisted of a sliding table on the tool carriage which was fed automatically and elevated by a screw. *(see right)*

MASSACHUSETTS ARMS CO., Chicopee Falls, MA

Builder, beginning in 1850 and continuing to the end of the Civil War, of "dial and milling engines, drill presses, tools and engines for gun work."

A Complete Lathe and Milling Machine.

For Bicycle, Clock, Gun, Electrical, Model, Tool, Instrument and light Machine Work.
WALLACE H. MANSFIELD,
71 Court Street, - New Haven, Conn.

MELLING-NORTHRUP CO., Jackson, MI

Formed about 1913 to make the Melling designed die-sinking machine *shown left*. Profiling and cherrying mechanisms were combined in the machine and oscillating cutters were used instead of rotating cutters in the cherrying device. In 1914, the firm was bought by Walcott & Wood Machine Co. and reorganized as the JACKSON MACHINE TOOL CO.

MERGENTHALER CO., O., Baltimore, MD

Builder of milling machines, patented November 14, 1899. The machine featured all geared drive and a very heavy overarm casting. As *shown right*, a vertical spindle attachment was mounted on the outboard end of the overarm.

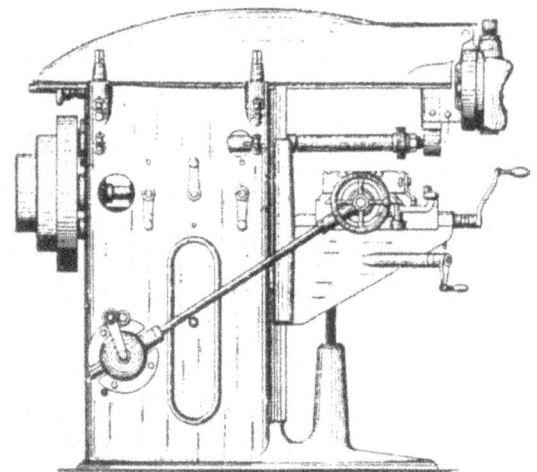

MERIDEN MACHINE CO., Meriden, CT

Formed in 1854 as a reorganization of OLIVER SNOW & CO. Gamaliel F. Snow was superintendent. The firm continued production of milling machines previously made by Oliver Snow & Co. Short lived, the business was reorganized as SNOW, BROOKS & CO. in 1855.

MERIDEN MFG. CO., Meriden, CT

Formed about 1865 as a reorganization of PARKER, SNOW & CO. Charles Parker was president. The firm continued production of milling machines previously made by Parker, Snow & Co. About 1868, Charles Parker reorganized the firm as PARKER BROTHERS.

MILLED SCREW & MACHINE CO., Sayre, PA

Formed in 1909 to make screw milling machines for production of lead and feed screws, elevator worms and other work of like nature. The machines (Fig.1) could mill screws up to 6" diameter and were made in lengths from 24" to 252". An improved version (Fig.2) was introduced in 1911 and was offered in the same sizes as the first model.

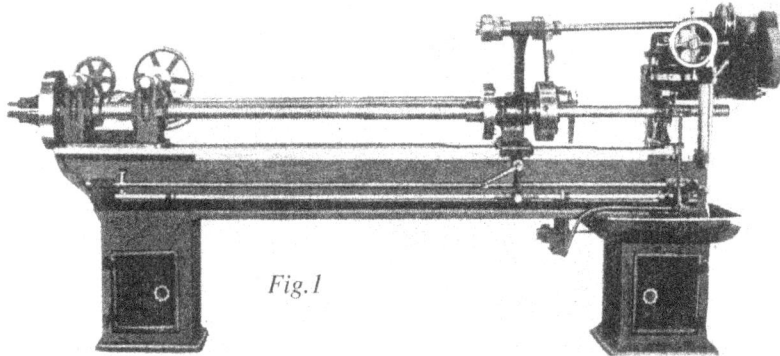

Fig.1

Fig.2

MILLER & CROWNSHIELD, Greenfield, MA

A partnership of Ernest Miller and Francis Crownshield formed in 1913 to make an automatic screw machine invented by Miller. A plain bench miller, with 4" x 12" table as *shown left,* was introduced in 1916.

MILWAUKEE MACHINE TOOL CO., Milwaukee, WI

Formed in 1888 by Frank Kempsmith (1848-1904) to make milling machines. Short-lived, the firm reorganized as the KEMPSMITH MACHINE TOOL CO. in 1889.

MORRIS MACHINE TOOL CO., Cincinnati, OH

Formed in 1910 as a reorganization of the John B. Morris Foundry Co. Engine lathes formed the bulk of the firm's production. A bench miller, *shown right*, was introduced in 1916. Vertical travel of 5" was in the head which also contained an overarm, an unusual feature for small milling machines.

BENCH MILLER

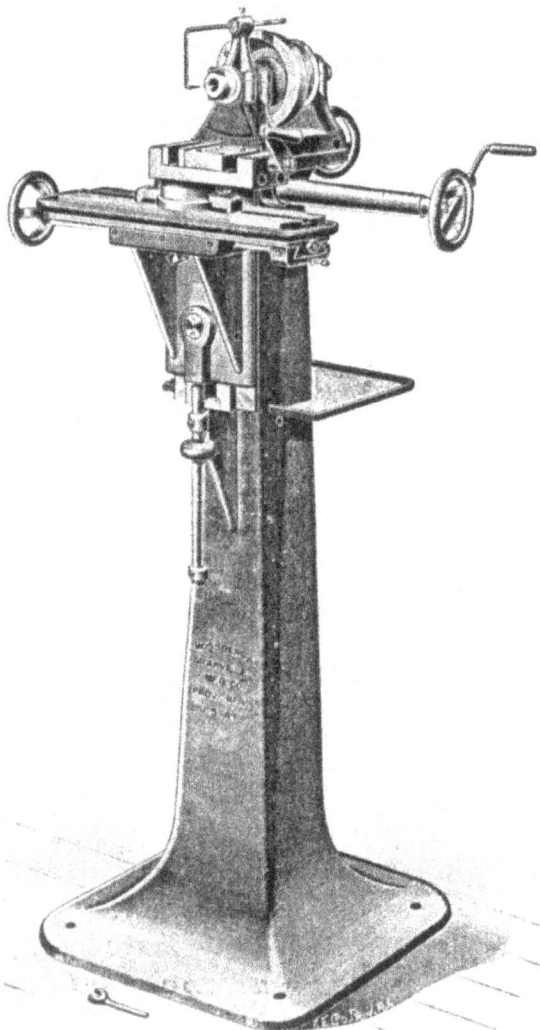

MOSSBERG & GRANVILLE MFG. CO., Providence, RI

Formed in 1896 by Frank Mossberg who had reorganized the Mossberg Mfg. Co. Primary products were presses, wire drawing machines and other metal forming machinery for the jewelry and silversmiths' trades. Production of these machines continued until 1903 when the firm reorganized as the Standard Machinery Co.

In 1897, the firm introduced a small, die making, milling machine for sale to their press users. Table travel was 12½", vertical travel 8½" and cross travel 3¾"; the head was indexable to 90 degrees.

NATIONAL MACHINE & TOOL WORKS,

Rockford, IL

Formed in 1909 as a reorganization of DALIN BROTHERS when Edwin Cedarleaf and O.J. Sundstrand bought control. The firm continued production of small milling machines begun by Dalin Brothers. Very short lived, the company became the ROCKFORD MILLING MACHINE CO. in 1910.

The Rockford Miller

Combination Screw and Lever Feed is a feature of this machine which, in addition to the line of attachments furnished, puts it among the most useful and convenient tools of its class. It is adapted for both vertical and horizontal milling, has an attachment for slotting, and an overhanging arm of great stiffness and capacity. If you have die work, slotting or light or heavy milling, this is the machine you need.

Write for full description.

NATIONAL MACHINE AND TOOL WORKS Rockford, Ill., U. S. A.

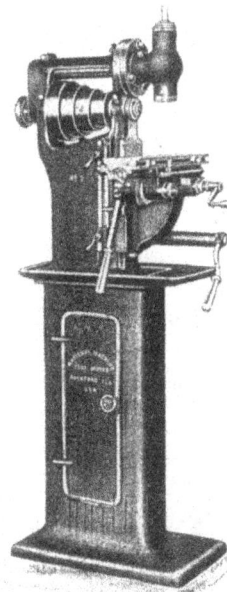

NATIONAL TRANSIT PUMP & MACHINE CO.,

Oil City, PA

A maker of pumps and other oil field machinery, the firm also built manufacturing milling machines during World War I. Although it looks very much like a knee type machine, it is actually closest in design to the Lincoln miller. As *shown at left*, the vertical motion is in the spindle head with the table and cross motion supported on a solid base. It was fitted with a variable speed electric drive.

NEW YORK STEAM ENGINE WORKS,

New York, NY and Worcester, MA

Formed about 1860 to make steam engines, the firm began the manufacture of machine tools in 1862 when it purchased the assets of THAYER, HOUGHTON & CO. Production of the Thayer, Houghton line of machine tools, including milling machines as *shown at right*, continued under the management of Alexander Thayer (1812-1895). Thayer left in 1864 to join the New Haven Mfg. Co. Alfred B. Couch (1829-1888) was then appointed superintendent and general manager. The Worcester factory was closed in 1871 and production moved to its Passiac, NJ, factory. Machine tools were produced until the factory closed in 1877.

NEWTON & COX, Dunkirk, NY, later Cleveland, OH

A partnership of Charles C. Newton (1846-1906) and Jacob D. Cox, formed July 27, 1876, to make twist drills and milling machines. The firm moved to Cleveland, OH, in September, 1876. Milling machine production began in 1878 and included hand milling machines (Fig.1), small power milling machines in two sizes (Figs.2-3), heavier milling machines with moveable overarms (Fig.4) and solid overarms (Fig.5), vertical milling machines for die sinking (Fig.6), and Lincoln type milling machines (Fig.7). In 1880 Newton sold out to Cox who reorganized as COX & PRENTISS. Newton moved to Philadelphia, PA, where he formed the NEWTON MACHINE TOOL WORKS. *(illustrations continued next page)*

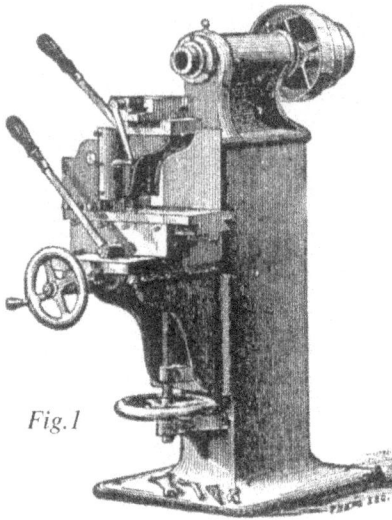

Fig.1

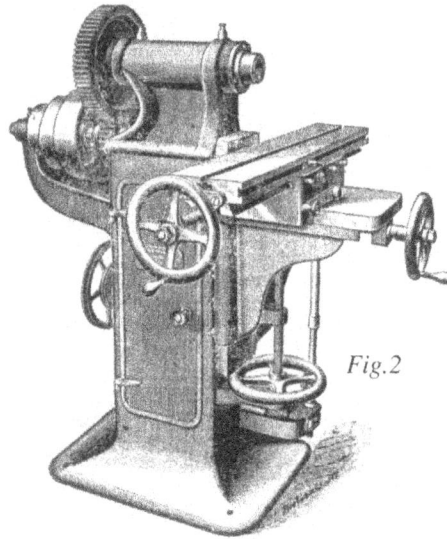

Fig.2

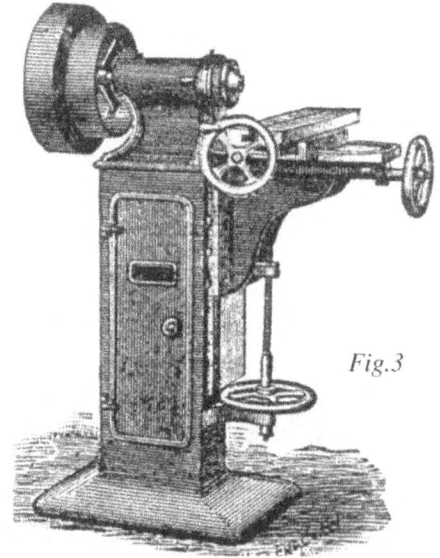

Fig.3

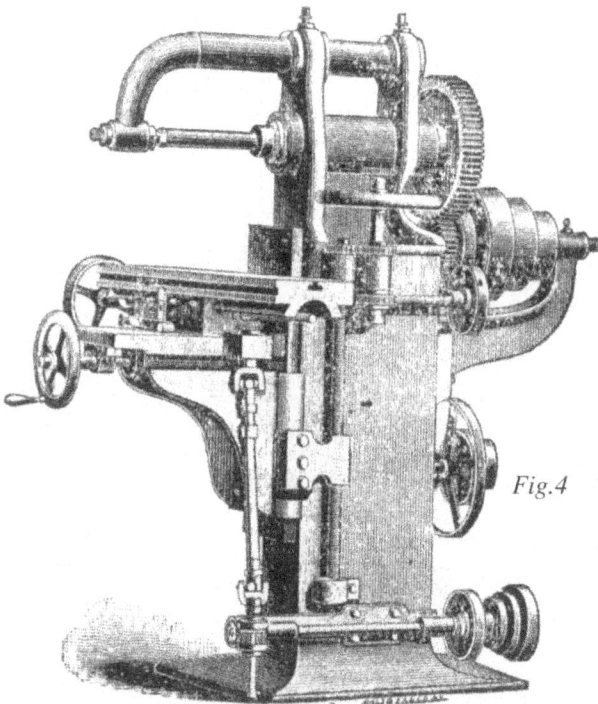

Fig.4

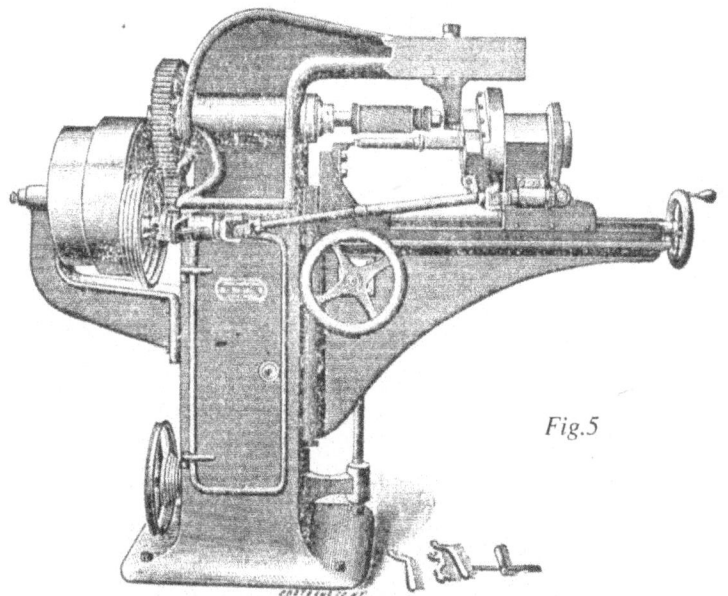

Fig.5

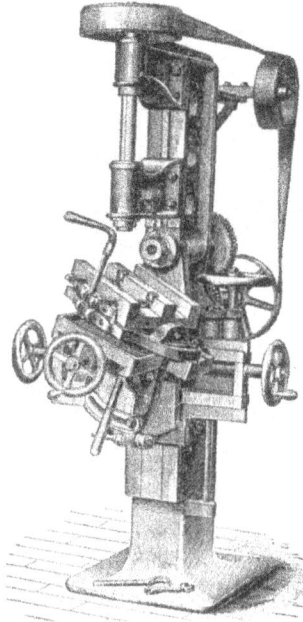

Fig.6

Fig.7

NEWTON MACHINE TOOL WORKS, Philadelphia, PA

Formed in 1880 by Charles C. Newton (1846-1906) after the partnership of NEWTON & COX was dissolved. Newton operated as a sole proprietor until incorporating July 14, 1897, and served thereafter as president and treasurer until his death in 1906.

First production was a continuation of milling machines previously made by Newton & Cox and included column milling machines (Fig.1), with 34" tables, 5" cross travel, and 12" vertical travel, priced at $350 or $375 with the optional overarm; smaller size column milling machines (Fig.2); hand milling machines (Fig.3) weighing 940 pounds and priced at $275; and combined milling and die sinking machines (Fig.4). Gun milling machines were introduced in 1881 in No. 1 (Fig.5) and No. 2 sizes, with a New Pattern No. 3 size (Fig.6) added in 1884. *(continued on next six pages)*

COLUMN MILLING MACHINE.

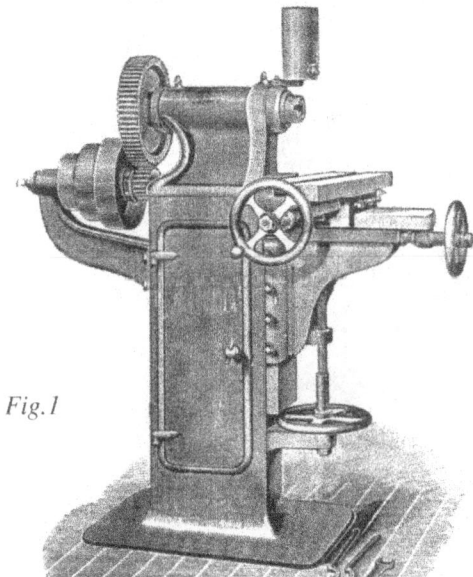

Fig.1

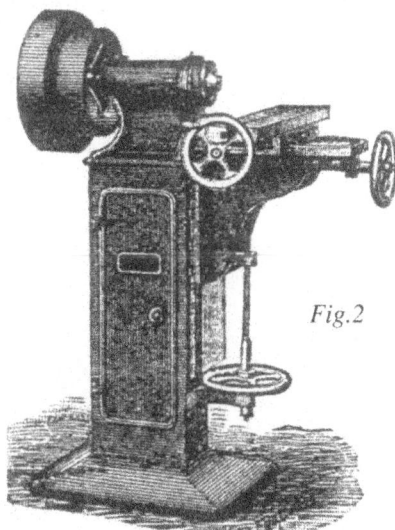

Fig.2

HAND MILLING MACHINE

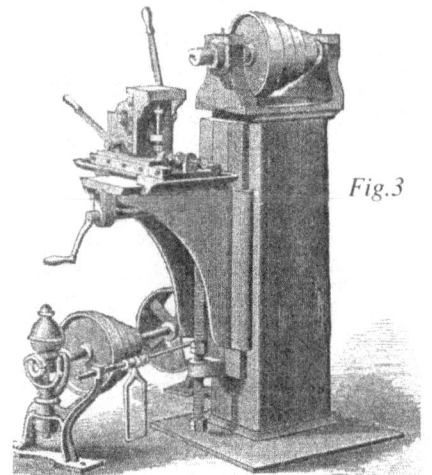

Fig.3

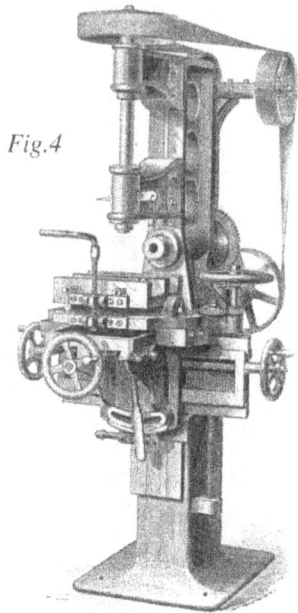

Fig.4

Fig.5

Fig.6

COMBINED MILLING AND DIE SINKING MACHINE.—

NEW GUN MILLING MACHINE.

NEW PATTERN MILLING MACHINE.

Other 1884 production included nut facing machines (Fig.7); double headed nut and bolt head millers (Fig.8), weighing 2,000 pounds and priced at $790; and locomotive rod millers (Fig.9) with 18" x 90" tables, weighing 8,000 pounds and priced at $1175. New machines offered in 1885 included a new slabbing and key-seat milling machine (Fig.10) built in three sizes, No. 1 with 9" x 48" table, No. 2 with 13" x 72" table, and No. 3 with 18" x 96" table; and vertical milling machines also built in three sizes, No. 1 (Fig.11) with 30" diameter table, No. 2 (Fig.12) with 32" table weighing 12,000 pounds, and No. 3 (Fig.13) with 36" table, weighing 16,000 pounds.

DOUBLE HEADED NUT AND BOLT HEAD MILLER.

Fig.7

Fig.8

-Nut-facing machine.

143

Fig.9

Fig.10

NEWTON'S SLABBING AND KEY-SEAT MILLING MACHINE.

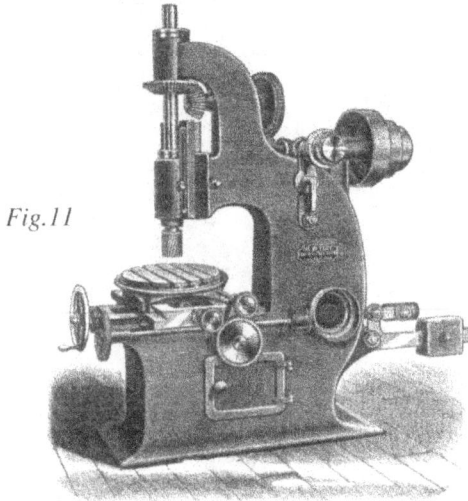

Fig.11

NEW VERTICAL MILLING MACHINE.

No. 2 VERTICAL MILLING MACHINE.

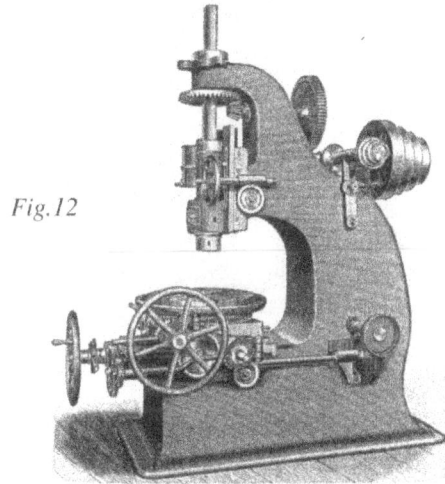

Fig.12

No. 3 VERTICAL MILLING MACHINE.

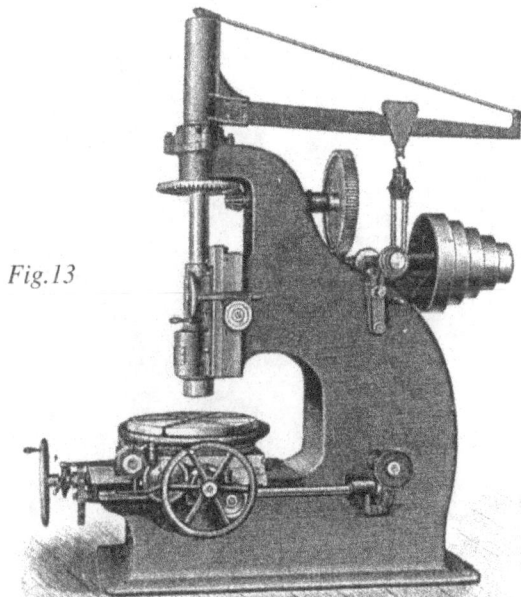

Fig.13

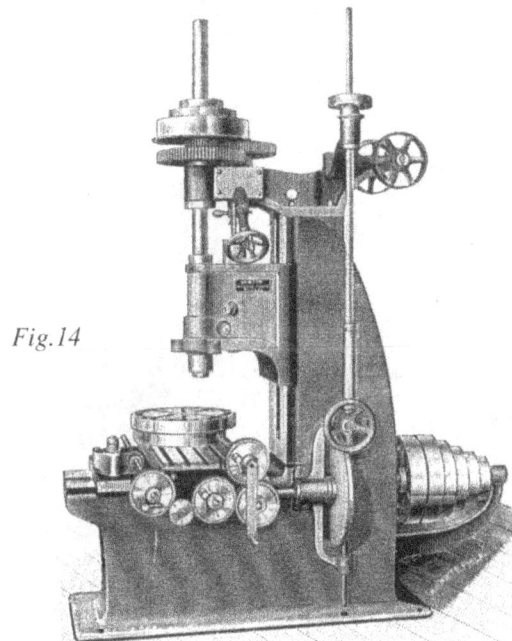

Fig.14

New design vertical milling machines were offered by 1893 in No. 1 size with 16" to 22" tables, No. 2 size (Fig.14) with 25" to 32" tables, and No. 3 size with 30" to 42" tables. New heavy milling machines (Fig.15), built in five sizes, were introduced in 1896.

1901 production included: duplex milling machines in No. 1 size (Fig.16) with 8" x 33" tables, weighing 2,100 pounds, No. 1½ size (Fig.17) with 15" x 46" tables, weighing 4,500 pounds, No. 2 size (Fig.18) with 12" x 53" tables, weighing 5,200 pounds, and No. 2½ size (Fig.19) with 12" x 72" tables, weighing 10,680 pounds; plain milling machines in No. 2 size (Fig.20) with 16" x 48" tables, weighing 8,500 pounds. No. 3 size (Fig.21) with 30" x any desired length tables, No. 4 size (Fig.22) with 30" x any desired length tables, and No. 5 size (Fig.23) with 36" x any desired length tables; vertical slabbing machines in No. 2 size (Fig.24) with one or two heads, No. 3 size (Fig.25) with 30" x 96" tables, and No. 4 size (Fig.26) with one or two heads and 36" x 96" tables; and No. 2 size vertical milling machines in universal type with an overhanging arm supporting the cutter (Fig.27) and special (Fig.28) with 36" x 72" tables, weighing 13,000 pounds.

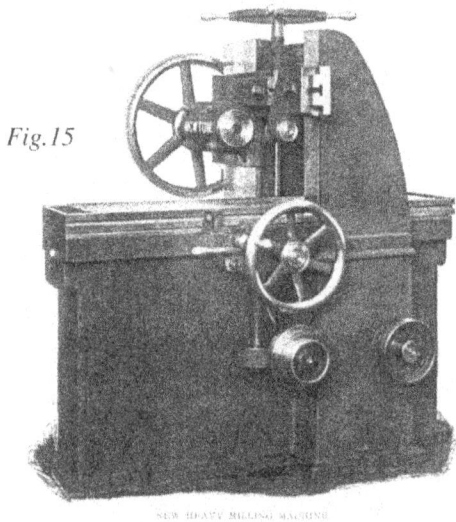

Fig.15

No. I DUPLEX MILLING MACHINE.

Fig.16

No. I½ DUPLEX MILLING MACHINE.

Fig.17

No. 2 DUPLEX MILLING MACHINE.

Fig.18

No. 2½ DUPLEX MILLING MACHINE.

Fig.19

No. 2 PLAIN MILLING MACHINE.

Fig.20

No. 3 PLAIN MILLING MACHINE.

Fig.21

No. 4 PLAIN MILLING MACHINE.

Fig.22

No. 5 PLAIN MILLING MACHINE.

Fig.23

No. 2 VERTICAL SLABBING MACHINE.

Fig.24

No. 3 VERTICAL SLABBING MACHINE.

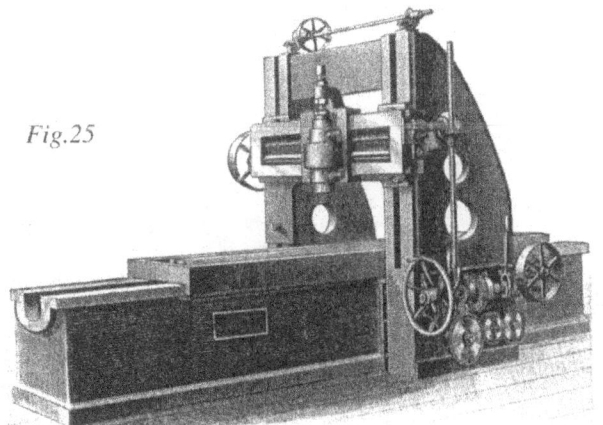

Fig.25

No. 4 VERTICAL SLABBING MACHINE.

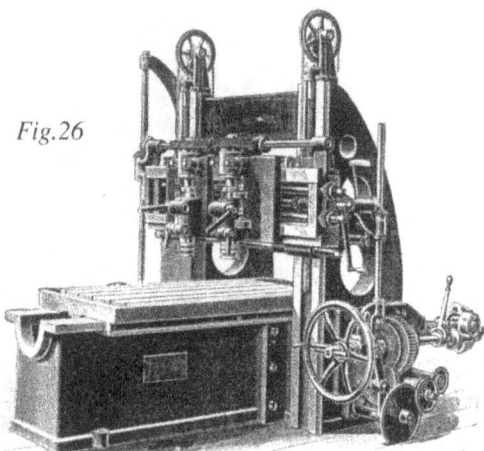

Fig.26

No. 2 VERTICAL UNIVERSAL MILLING MACHINE.

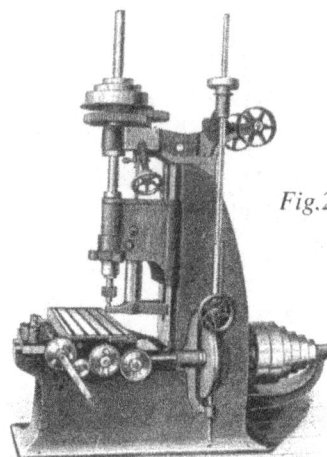

Fig.27

SPECIAL No. 2 VERTICAL MILLING MACHINE.

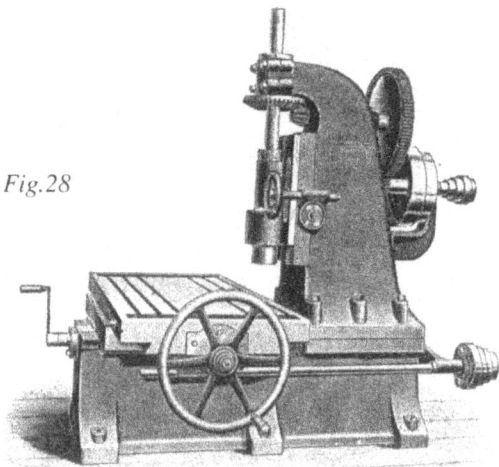

Fig.28

Fig.29

THE NEWTON MILLING AND DRILLING EQUIPMENT FOR STEEL TRUCKS.

Motor drive machines were also offered in 1901 and included 7.5 HP plain millers (Fig.29), 7.5 HP double spindle milling machines (Fig.30), and vertical spindle mills (Fig.31) for use with cutters up to 30" diameter. Heavy slab milling machines (Fig.32) continued in production through World War I. Thread milling machines (Fig.33) were introduced in 1916. The Newton Machine Tool Works was absorbed by the Consolidated Machine Tool Corp. of American in 1922. *(illustrations continued next page)*

Fig.30

Double Spindle Milling Machine.

Fig.31

Fig.32

Fig.33

NEWTON THREAD MILLING MACHINE

NICHOLSON & WATERMAN, Providence, RI

A partnership of Stephen Nicholson and his brother-in-law Stephen Waterman formed in 1887. Nicholson was the son of William T. Nicholson, founder of the Nicholson File Co. Short lived, the firm appears to have gone out of business about 1892. Products included engine lathes, drilling machines, and a combination lathe, milling machine and boring machine as shown below. When used as a milling machine it is, in effect, a Lincoln miller with short table travel.

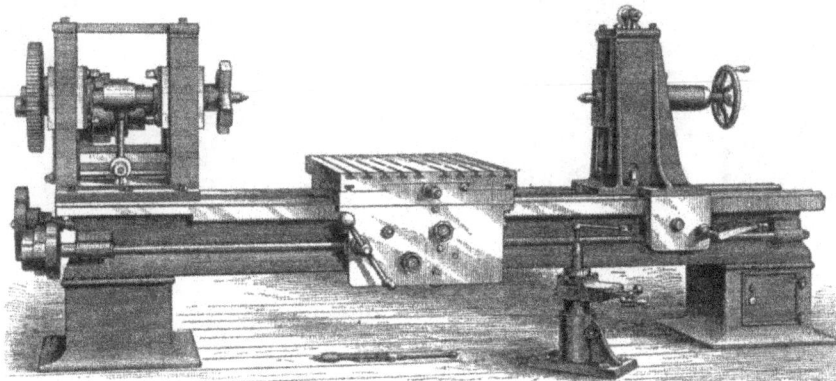

HORIZONTAL BORING, TURNING AND MILLING MACHINE.—NICHOLSON & WATERMAN.

NICHOLSON, WILLIAM T., Providence, RI

Builder, from 1860 to 1863, of "hand and power feed milling machines, furnished for $300 per pair". Nicholson later went on to found the Nicholson File Co.

NILES & CO., Cincinnati, OH

later **NILES TOOL WORKS,** Hamilton, OH

A partnership of James Niles (1809-1881) and his brother Jonathon Niles (1804-1878) formed in 1845. Early products included steam engines and locomotives. The brothers began making machine tools about 1859. In 1869, the brothers sold out to Alexander Gordon (1839-1910), George A. Gray, Jr. (1839-1905) and James W. Gaff. The partnership of Gaff, Gordon & Gray was formed which operated the company as the Niles Tool Works. The new firm specialized in large machine tools for the railroad industry. The factory was moved to Hamilton,

OH, in 1873 and the firm incorporated in 1874. Gray left in 1877 to form G.A. Gray, Jr. & Co. Gaff died in 1879; his son, James W. Gaff, Jr. succeeded his father as president. George T. Reiss (1849- 1915) was appointed chief engineer in 1879 and was responsible for machine design for Niles and, later, Niles-Bement-Pond, until his death in 1915. Alexander Gordon became president when the firm was reorganized in 1892, serving until August 15, 1899, when the Niles Tool Works merged with BEMENT, MILES & CO. and the POND MACHINE TOOL CO. to form the Niles-Bement-Pond Co. Niles continued operation as a division of Niles-Bement-Pond Co.

Large planers, boring mills and lathes formed the bulk of production but some milling machines were produced, including large planer mills in horizontal spindle (Fig.1) to mill 12' long and 36" wide and vertical spindle (Fig.2) to mill 12' long and 36" wide, made in 1895. Transverse facing and milling machines (Fig.3) weighing 23,000 pounds; and horizontal duplex milling machines (Fig.4), were introduced in 1897. 1904 production included circular milling machines (Fig.5), for milling circular work such as large gears; 9-spindle milling machines (Fig.6) for facing cast iron heads for water tube boilers; vertical milling machines (Fig.7) with 42" rotary tables; horizontal milling and facing machines (Fig.8) with 48" tables; and portable milling machines (Fig.9) designed for use on a floor plate. *(illustrations continued on next page)*

Fig.1

Fig.2

Fig.3

Fig.5

Fig.4

Fig.6

Fig.7

Fig.8

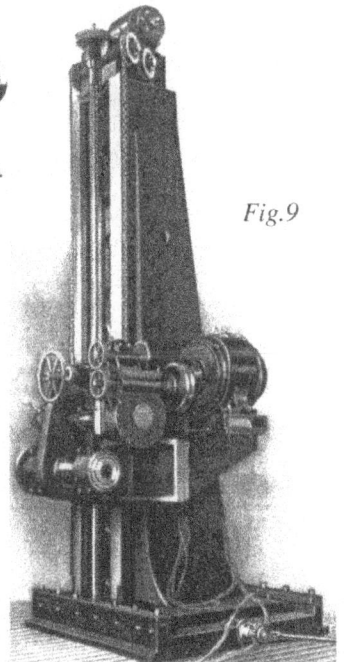

Fig.9

OESTERLEIN & CO., WILLIAM, Cincinnati, OH
later OESTERLEIN MACHINE CO., Cincinnati, OH

Founded in 1886 by William Oesterlein (1848-1920). The firm reorganized as the Oesterlein Machine Co. in 1898 with William Oesterlein serving as president until his death in 1920.

Milling machine production appears to have begun in 1898 with the introduction of a line of plain and universal millers. Plain millers included a 20 x 6½ x 7 inch size (Fig.1) weighing 1,450 pounds; 24 x 6½ x 18 inch size (Fig.2), weighing 1,800 pounds; and 34 x 9 x 10 inch size (Fig.3), weighing 2,800 pounds. Universal millers included a 20 x 6½ x 17 inch size (Fig.4) with dividing head, weighing 1,800 pounds; and 24 x 7½ x 18 inch size (Fig.5) with dividing head, weighing 2,400 pounds. A combination vertical spindle milling (Fig.6) and rack cutting (Fig.7) attachment was offered with the machines. *(continued on next two pages)*

20 X 6½ X 7 INCH.

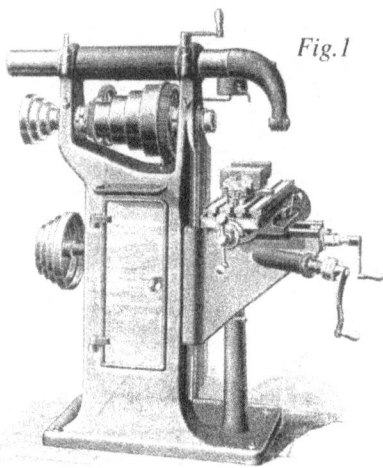

Fig.1

24 X 6½ X 18 INCHES.

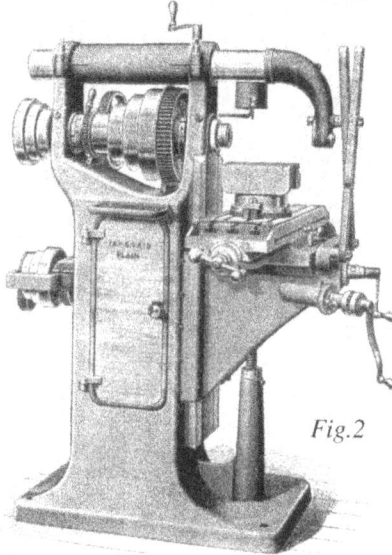

Fig.2

34 X 9 X 10 INCHES.

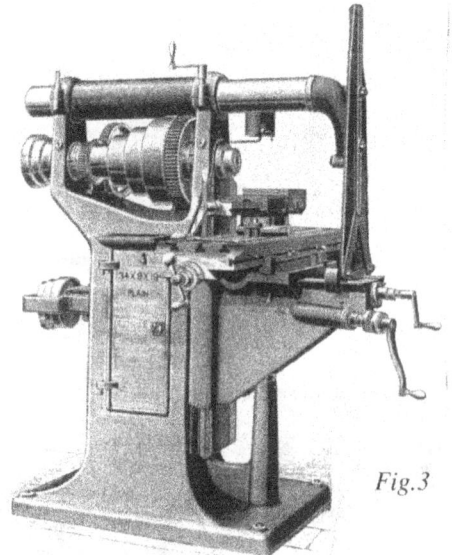

Fig.3

20 x 6½ x 17 INCH.

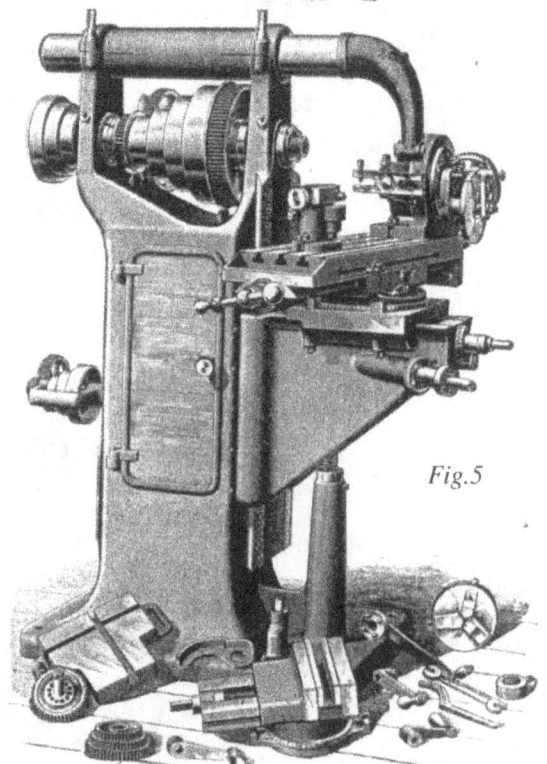

Fig.4

Fig.5

Fig.6

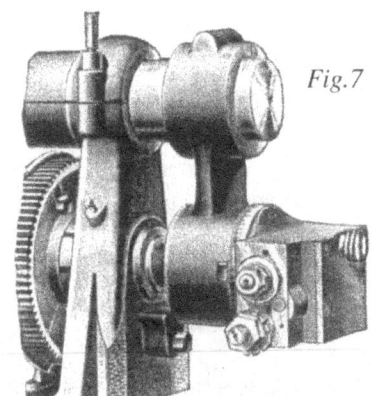

Fig.7

(Showing Rack-cutting Spindle in use).

Fig.8

THE OESTERLEIN UNIVERSAL MILLING MACHINE.

Fig.9

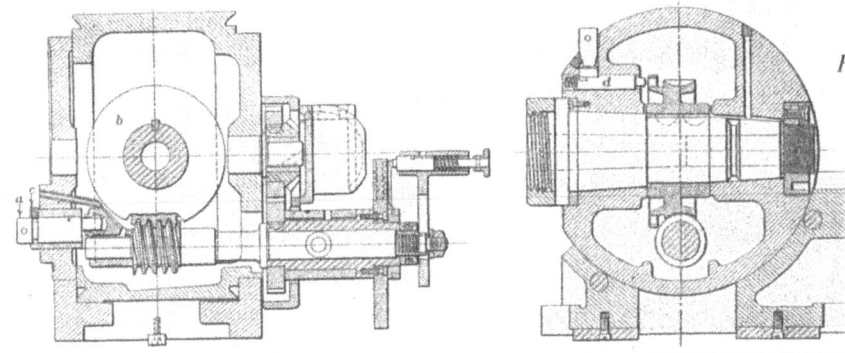

Fig.10

Fig. 2.—Sectional Elevations of the Universal Dividing Head.

Fig.11

The New 20x17x7½ In. Universal Milling Machine Built by the Oesterlein Machine Company, Cincinnati, Ohio

Fig.12

A new line of universal milling machines (Fig.8) was introduced in 1903 with another redesign, designated the No. 25 (Fig.9) introduced in 1910, along with a new universal dividing head (Fig.10). A further improved universal milling machine (Fig.11) and a new design plain miller (Fig.12) were brought out in 1913. Power traverse for the knee motion was added in 1915. Constant speed models (Fig.13) were available about 1920.

Continuous milling machines (Fig.14) with the table set at a 15 degree angle to facilitate loading and unloading were introduced in 1919.

Fig.13

Fig.14

OWEN MACHINE TOOL CO., Springfield, OH

Founded in 1893 by William H. Owen (?-1906) after leaving the Springfield Machine Tool Co. where he was co-founder.

Milling machine production began by 1900 with the No. 3 plain (Fig.1) equipped with 33" of table travel and 8½" of cross. No. 2 universal milling machines (Fig.2) were offered by 1904 and No. 2-B (Fig.3) and 3-B plain millers (Fig.4) by 1907. Duplex milling machines (Fig.5) were marketed by 1908. The company appears to have gone out of business about 1914.
(illustrations continued next page)

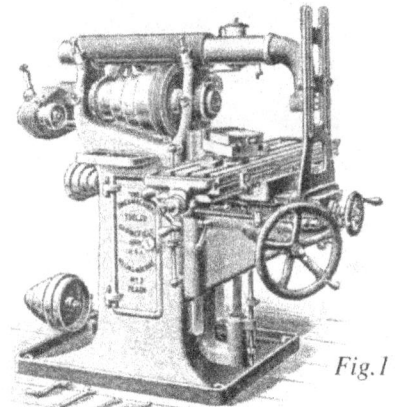
Fig.1

Fig.2

Fig.3

Fig.4

Fig.5

SPECIFICATIONS

Table has an automatic feed of....................	26¼ in.
Vertical adjustment of spindles..................	6 in.
Max. width of piece that can be milled..........	26 in.
Table dimensions........	44x7½ or 44x9 in.
Drive Pulley, diameter......................	16 in.
Width of Belt	3 in.
Spindle, taper hole........	No. 10 B. & S.
Spindle hole...............................	1¼ in.
Feed, number of changes	6
Feed, per 1 revolution of spindle.......	.016 in. to 120 in.
Feed belt	1¼ in.
Counter Pulleys, diameter...................	14 in.
Counter Pulleys, face.......................	4⅝ in.
Net weight...............................	2450 lbs.
Gross weight, domestic shipment	2900 lbs.

PARKER BROTHERS, Meriden, CT

A partnership of Charles Parker (1809-1902) and Edmund Parker, formed in 1868 as a reorganization of the MERIDEN MFG. CO., which had been a builder of milling machines. Parker Brothers mainly manufactured a wide variety of cast iron products such as coffee mills, vises, etc. Production of milling machines, probably of the type made by the Meriden Mfg. Co. and predecessor firms during the Civil War, was continued. One such, photographed still in production at Parker Brothers in 1913, is *shown below*. Other machine tool production began in 1869 with the introduction of presses and continued past 1882 when they began making planers.

FRONT VIEW OF PARKER BROS. MILLER

REAR VIEW SHOWING MECHANISM

PARKER, SNOW, BROOKS & CO., Meriden, CT

later **PARKER, SNOW & CO.,** Meriden, CT

Formed in 1862 as a reorganization of SNOW, BROOKS & CO. when Charles Parker became the senior partner. The firm continued production of milling machines previously made by Snow, Brooks & Co. Brooks left the firm in 1863 and the firm became Parker, Snow & Co. In 1865, Parker reorganized as the MERIDEN MFG. CO.

PATENT ARMS MFG. CO., Paterson, NJ

Founded by Samuel Colt in 1836 to make a newly designed revolving pistol. Short lived, the company failed in 1842. During this short period, Colt employee William Ball designed and built a number of machine tools including index milling machines, screw machines and drill presses. When the firm's assets were sold at auction in 1845, the machinery brought in more than $6,000.

PEDRICK & AYER, Philadelphia, PA

later **PEDRICK & AYER CO.,** Philadelphia, PA

A partnership of Daniel W. Pedrick (1840-1926) and Henry C. Ayer, formed in 1881 to operate the L.B. Flanders Machine Works. In 1886, the partners dropped the Flanders name, reorganized as Pedrick & Ayer, and continued production of many of the Flanders products, which were mainly portable machine tools for locomotive repair shops. Production included portable milling machines for milling locomotive steam chests (Fig.1) which had been made by A.H. CAMPBELL from 1881 to 1884.

Heavy milling machines were introduced about 1887 when Pedrick & Ayer took over production of a universal milling machine previously made by DETRICK & HARVEY under Cosgrove's patent of August 7, 1883. The first model was a heavy universal milling machine (Fig.2) with 52" x 14" table, furnished with a vertical attachment (Fig.3) and COSGROVE'S universal vise (Fig.4). A No. 1 universal milling machine (Fig.5) was intro-

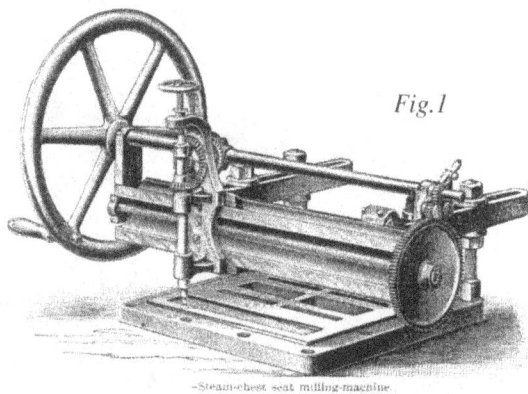

Fig.1

–Steam-chest seat milling-machine.

Fig.2

Fig.3

Fig.4

duced in January, 1889, and was offered with arbor supports (Fig.6); vertical milling attachments, mountable in several ways (Figs.7-10), for use with an index plate or universal vise; and boring attachments (Fig.11). *(continued on next two pages)*

Fig.5

NO. 1 UNIVERSAL MILLING MACHINE, BUILT BY PEDRICK & AYER.

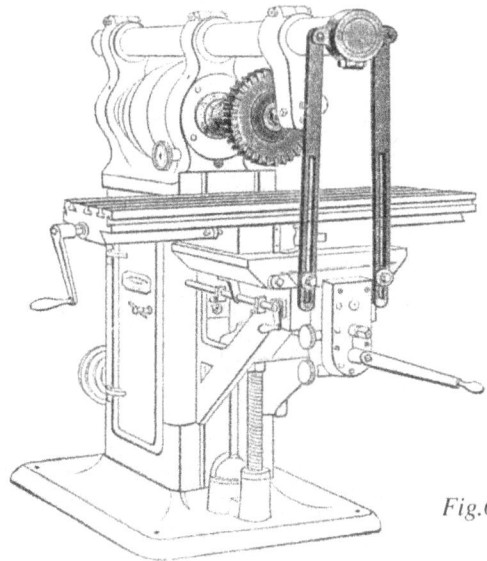

Fig.6

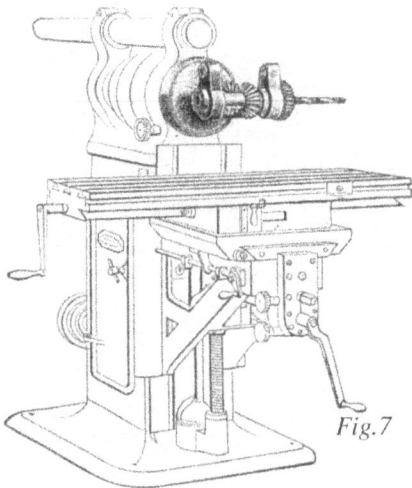

Fig.7

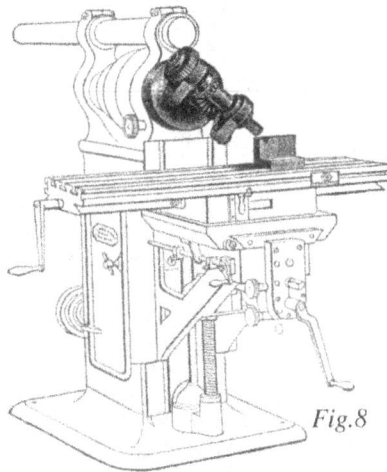

Fig.8

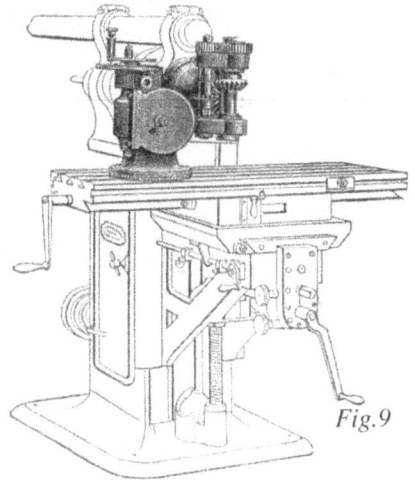

Fig.9

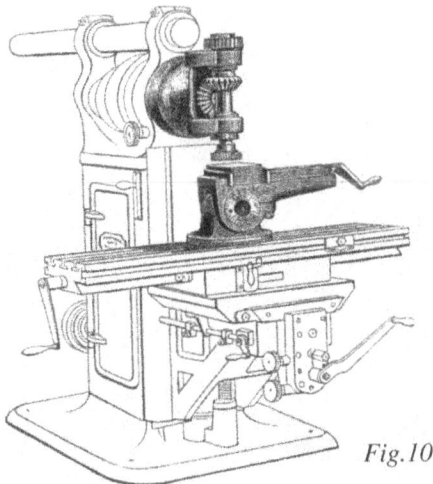

Fig.10

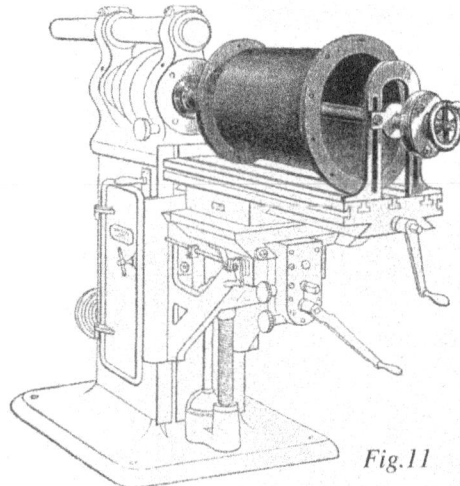

Fig.11

Heavy plain milling machines (Fig.12) with 48" x 14" tables were introduced in 1892. Universal (Fig.13) and vertical (Fig.14) index heads were introduced in 1892 along with heavy vertical milling machines with 48" x 14" tables (Fig.15). In 1894, the heavy universal milling machine was introduced in England, complete with a startling array of attachments (Fig.16). A duplex boring machine (Fig.17) clearly based on the plain milling machine design, was offered in 1890.

In 1894, the firm incorporated as the Pedrick & Ayer Co. with capital of $100,000. The machinery sales firm of Manning, Maxwell & Moore, which was to be the selling agent and business manager of the firm, was the majority stockholder. Milling machine production continued until at least 1902; the firm ceased operations in 1907.

Fig.12

Fig.13

Fig.14

Fig.15

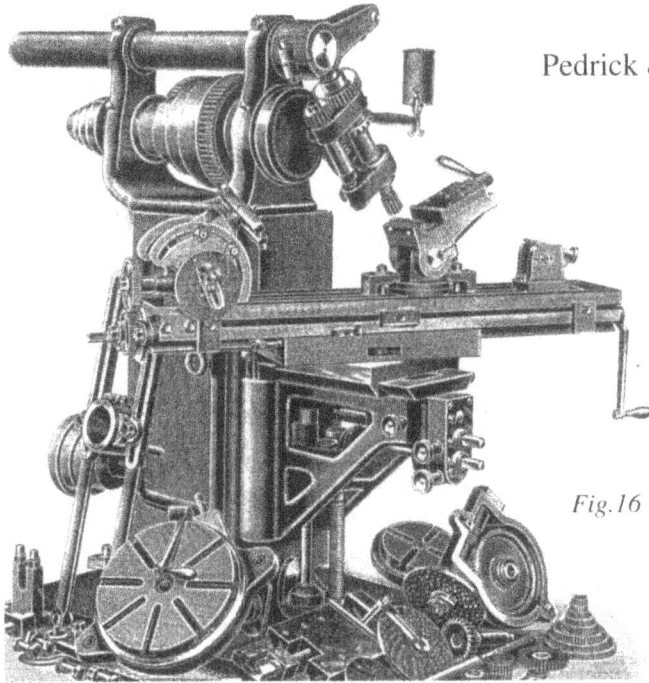

Pedrick & Ayer

Fig.16

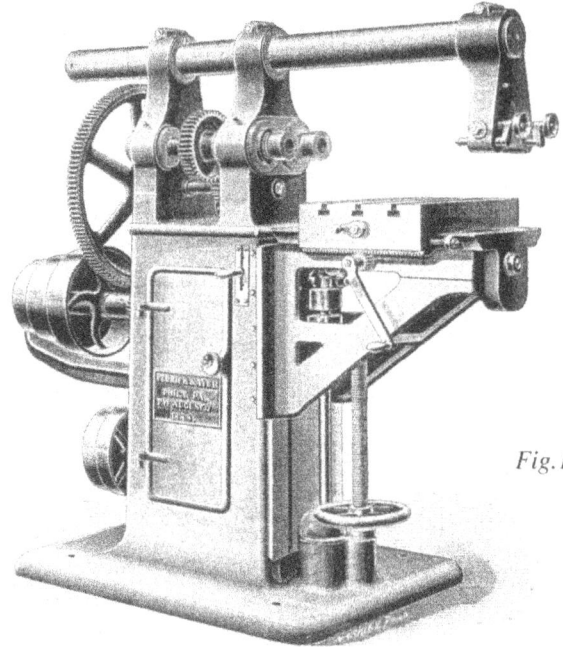

Fig.17

DUPLEX BORING MACHINE.

PHILADELPHIA MACHINE TOOL CO.,
Philadelphia, PA

Formed in 1896 as a reorganization of the HENDERSON MACHINE TOOL WORKS. In 1902, the firm merged with A. FALKENAU to form the Falkenau-Sinclair Machine Co. The only known milling machine production was a duplex vertical machine introduced in 1900 and *shown at right*. The machine was designed for simultaneously cutting keyways in both ends of a shaft.

THE PHILADELPHIA DUPLEX VERTICAL MILLING MACHINE

PHOENIX IRON WORKS, Hartford, CT

Operated by GEORGE S. LINCOLN & CO., a partnership of George S. Lincoln (1819-1894), and his brother Charles L. Lincoln (1825-1908), formed in 1846 to make a variety of large machine tools, but specializing in lathes, planers, and milling machines. The firm reorganized as LINCOLN & CO. when George S. Lincoln retired in 1885 and again as the PHOENIX IRON WORKS CO. when Lincoln's grandson, Charles L. Taylor took over in 1901.

The famous LINCOLN milling machine (Fig.1) was developed here in 1854 by Francis A. Pratt, later one of the founders of the PRATT & WHITNEY CO. The most notable improvement over earlier designs such as the HOWE miller was the use of screw feed for the table. The Lincoln miller became a generic type and was made by many different machine tool builders well into the 20th century. Index milling machines (Fig.2), also designed by Francis A. Pratt, were offered by 1861 when the Colt Patent Firearms Mfg. Co. placed an order for 100 at $235 each. *(illustrations on next page)*

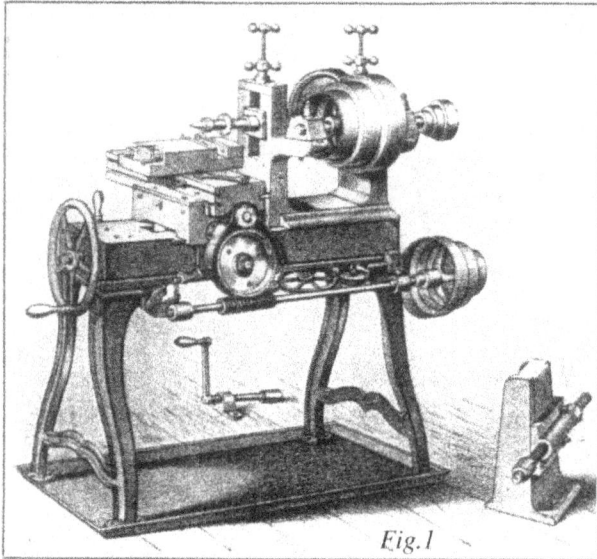

Fig.1

Fig.2

POND, D.W., Worcester, MA

later **POND MACHINE TOOL CO.,** Worcester, MA, later Plainfield, NJ

Founded by David W. Pond (1848-1897) in 1872 after leaving a partnership with his father Lucius W. Pond. Early production included engine lathes, vertical drilling machines and index milling machines.

The index milling machines (Fig.1) were offered in the LUCIUS W. POND 1874 catalog and were available with centers (Fig. 2 and vises (Fig.3) which mounted on the index arbor.

Pond bought out his father's creditors when L.W. POND failed in 1875, merged the firm with his own company and operated for a time as D.W. POND, SUCCESSOR TO L.W. POND. In 1882 Pond was the largest machine tool builder in Worcester, employing 240 men. The firm was incorporated February 1, 1883, as the POND MACHINE TOOL CO. with D.W. Pond as president. The company moved to Plainfield, NJ, in 1887 with Pond now serving as superintendent. He sold his interest to MANNING, MAXWELL & MOORE in 1894 and shot himself August 4, 1897. On August 15, 1899, the company merged with the NILES TOOL WORKS and BEMENT, MILES & CO. to form the NILES-BEMENT-POND CO. The POND MACHINE TOOL CO. continued to operate as a division.

Fig.1

Fig.2

Centres for Milling Machine.

Fig.3

Vise for Milling Machine.

INDEX MILLING MACHINE.

POND, L.W., Worcester, MA

Formed in 1856 by Lucius W. Pond (1826-1889) as a successor to SAMUEL FLAGG & CO., where Pond had been a partner. By 1858, Pond was offering "planers from 2 to 36 feet, engine lathes, and upright drills." In 1867, Pond offered index milling machines and gear cutting machines, *at right* . Pond took his son, David W. Pond, as a partner about 1869 but David Pond left to form his own company in 1872. L.W. Pond's business was taken over by his creditors in 1875 when he pleaded guilty to three counts of forgery and was sentenced to 15 years in prison. His son bought the firm from the creditors and merged it into D.W. POND which continued production of the index milling machines and gear cutting machines.

GEAR CUTTER.

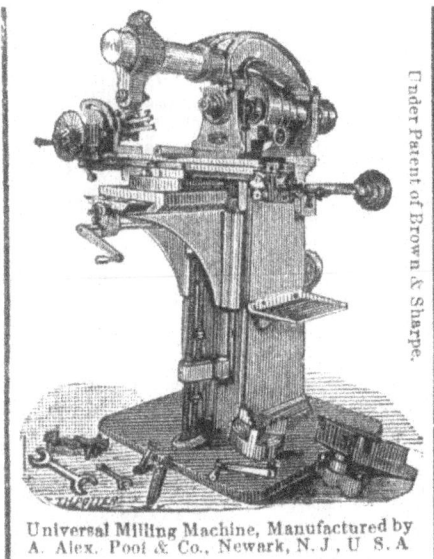

Under Patent of Brown & Sharpe.

Universal Milling Machine, Manufactured by
A. Alex. Pool & Co., Newark, N.J., U S.A.

POOL & CO., A. ALEX, Newark, NJ

Founded 1873 by Alex A. Pool to make plain and universal milling machines, engine lathes and bench lathes. The universal milling machine, *shown at left*, is clearly a Brown & Sharpe copy made, as the ad notes, "Under patent of Brown & Sharpe". In 1880, Pool advertised "combined lathes and milling machines for tool makers' use". The firm ceased operation in 1883.

POTTER & JOHNSTON MACHINE CO.,
Pawtucket, RI

Formed in April, 1893, by James C. Potter (?-1925) and John Johnston to make manufacturing lathes and shapers. Johnston had previously been chief draftsman of the PRATT & WHITNEY CO. The only known milling machine produced by Potter & Johnston is a very advanced two-station machine designed by James C. Potter in 1914. As *shown right,* it was equipped with twin tables that rotated into position. The table not under the cutter could be loaded and unloaded by the operator while the machine continued in operation. This feature, now known as pallet changing, was introduced as something new and wonderful in the first machining centers when developed in the late 1950s.

PRATT, WHITNEY & CO., Hartford, CT
later PRATT & WHITNEY CO., Hartford, CT

A partnership of Francis A. Pratt (1827-1902) and Amos Whitney (1832-1920), formed in 1860 to make silk machinery, and machine tools. Both had previously worked at the PHOENIX IRON WORKS, operated by George S. Lincoln & Co., where Pratt was superintendent and Whitney a contractor. Pratt, while working for Phoenix in 1855, had developed the famous Lincoln milling machine.

The firm incorporated as the Pratt & Whitney Co. in 1869. Employment in 1883 was 675 hands, increasing to 1,160 in 1896. Pratt served as president until his retirement in 1898; Whitney was superintendent, becoming president in 1898. In January, 1901, Whitney sold out to the Niles-Bement-Pond Co. which continued operation of the Pratt & Whitney Co. as a division.

Production of milling machines, along with a variety of other machine tools, had begun by 1868 when the firm ran one of their first ads (Fig.1). By 1876, the firm offered Lincoln type milling machines (Fig.2) of the type developed by Pratt in 1855, vertical die sinking millers (Fig.3), hand milling machines (Fig.4) with 6" table travel, 2" cross travel and 2" vertical travel, No. 1 column milling machines (Fig.5), and index milling machines (Fig.6). Profiling machines (Fig.7) were introduced by 1880.

Fig.1

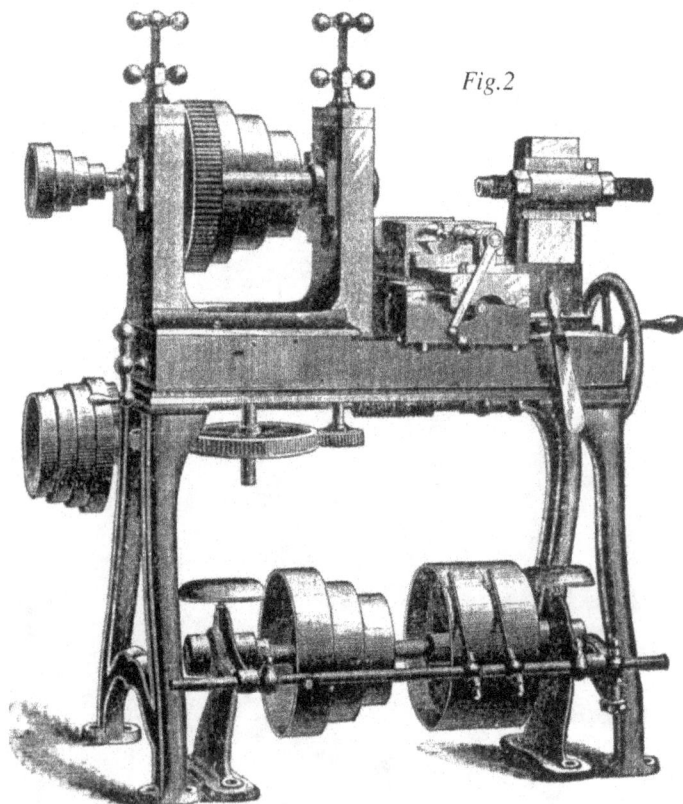

SUPERIOR TOOLS.

ENGINE LATHES, 15-inch swing, turn 2½, 3½, 4½ feet, with screw-gear, with or without taper attachment. One ditto, 27-inch swing, with screw-gear, cross-feed, and taper attachment, turns 10 feet. HAND LATHES, 13-inches swing, 5, 6, and 8 feet beds. CRANK PLANERS, stroke 8 inches, bed traverse 16 inches. GEAR PLANERS, 20 by 20 inches, by 4½ and 6 feet. PLANER VISES and CENTERS and MILLING MACHINES, ready for delivery.

16 by 16 inch, 24 by 24 inch, 30 by 30 inch PLANERS, various lengths; 19, 21, and 24 inch LATHES; ONE, THREE, AND FOUR SPINDLE DRILLS; SCREW MACHINES; HYDRAULIC ENGINES, etc., finishing.

For engravings and terms address
PRATT, WHITNEY & CO.,
25 and 1   Hartford, Conn.

Fig.2

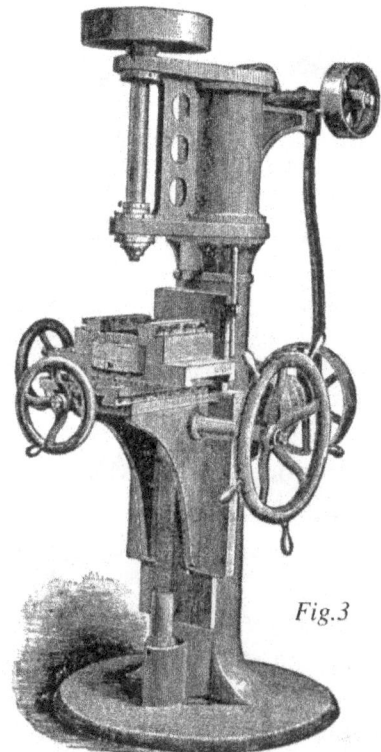

Fig.3

161

Fig.4

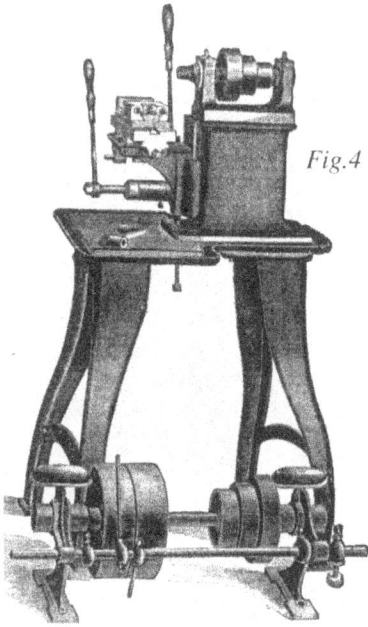

Fig.4

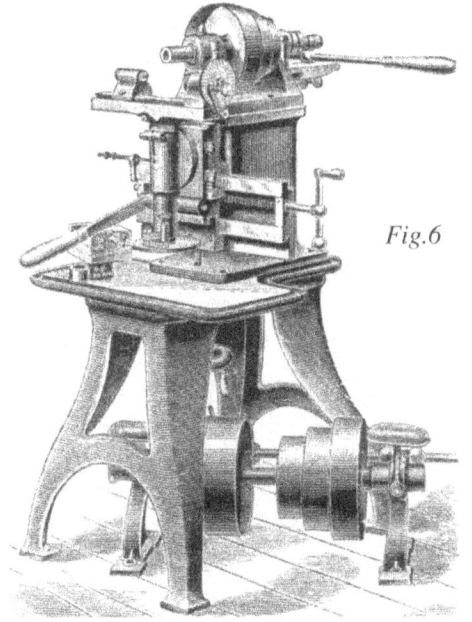

Fig.6

Fig.5

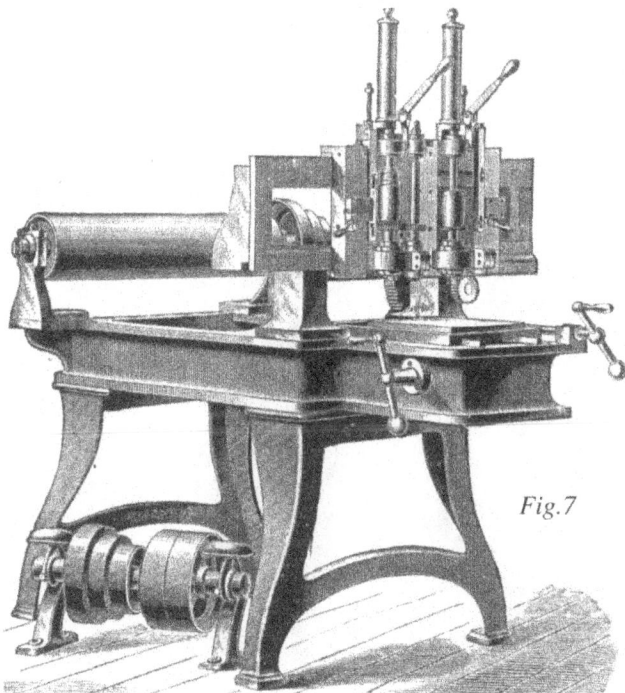

Fig.7

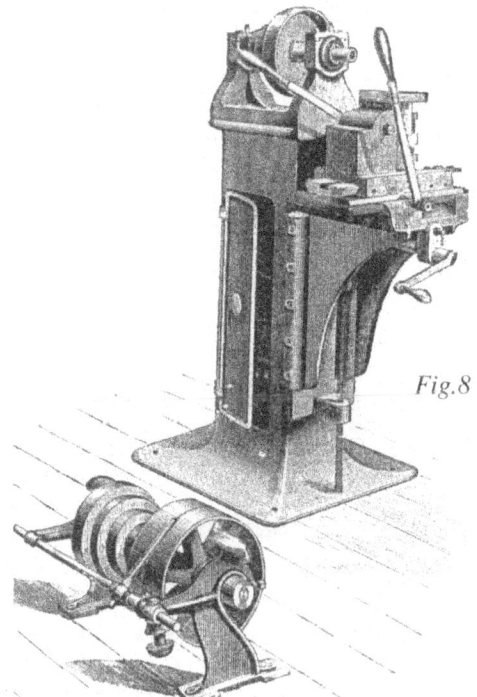

Fig.8

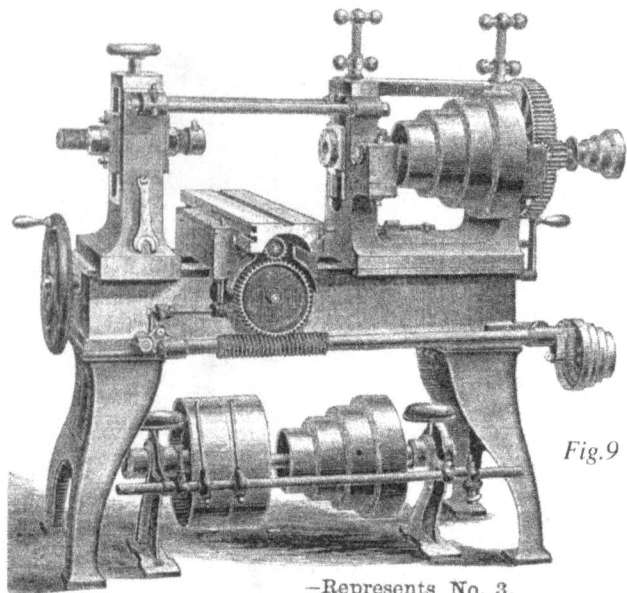

Fig.9

—Represents No. 3.

Fig.10

New Shuttle Race Milling Machine.

1884 production included hand milling machines (Fig.8); improved Lincoln millers in No. 2, No. 3 (Fig.9) weighing 2,150 pounds, and No. 4 sizes weighing 3,100 pounds; shuttle race milling machines (Fig.10) for milling the shuttle races in the beds of sewing machines; double spindle milling machines (Fig.11) designed for milling the faces and arms of sewing machines, while taking rough and finish cuts in one pass. The machine weighed 1,500 pounds and could take parts up to 20" in length.

Spline milling machines (Fig.12) for shafts up to 6" diameter, and the John J. Grant-designed double-column milling machine (Fig.13) with 16" table travel and weighing 2,000 pounds, were introduced in 1889.

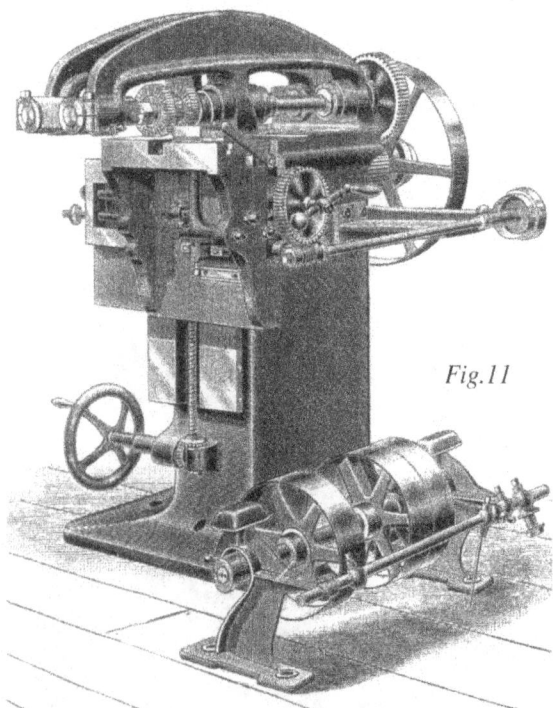

Fig.11

New Double Spindle Milling Machine.

Fig.12

Spline Milling Machine.

Fig.13

The Grant Double-Column Milling Machine.

163

1892 production included circular milling machines (Fig.14) for interior or exterior milling of large circular workpieces such as gear blanks, planer mills (Fig.15) designed by John E. Sweet and patented August 11, 1896; knee type mills in No. 0 size with 10" table travel, 4" cross travel and 10" vertical travel, and No. 1 (Fig.16) size with 15" table travel, 5" cross travel and 15" vertical travel, equipped with a new overarm design; and double-head milling machines (Fig.17) weighing 21,800 pounds. Valve milling machines (Fig18) were introduced in 1893, and four-spindle milling machines (Fig.19) in 1894.

Fig.14

CIRCULAR MILLING MACHINE, BUILT BY THE PRATT & WHITNEY COMPANY.

Fig.15

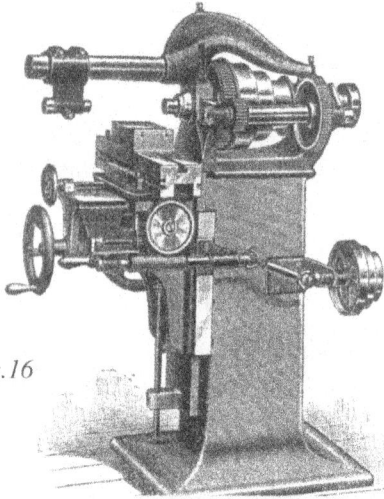

Fig.16

Fig.17

DOUBLE-HEAD MILLING MACHINE, MADE BY THE PRATT & WHITNEY COMPANY.

Fig.18

VALVE MILLING MACHINE.

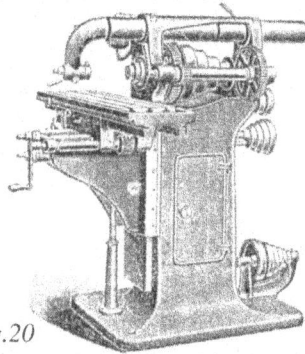

Fig.20

COLUMN MILLING MACHINES,

IN FOUR SIZES.

No. 0 and No. 1 with and without overhanging arm and back-gear. No. 2 and No. 3, shown in cut, have reversible feed. All have self-acting screw feed and stop motion, and tables have patent adjustable stop mechanism which prevents breakage of teeth when the work is run up to the cutters. Their stiffness and power fit them to do their work rapidly and smoothly, with long life to themselves and the mills.

Ask for Machinists' Catalogue.

THE PRATT & WHITNEY CO.,

HARTFORD, CONN., U. S. A.

NEW YORK : 123 Liberty St.   BOSTON : 144 Pearl St.
                CHICAGO ; 42 South Clinton St.
LONDON, ENGLAND : Buck & Hickman, 280 Whitechapel Road.
PARIS, FRANCE ; Fenwick Freres & Co., 21 Rue Martel.

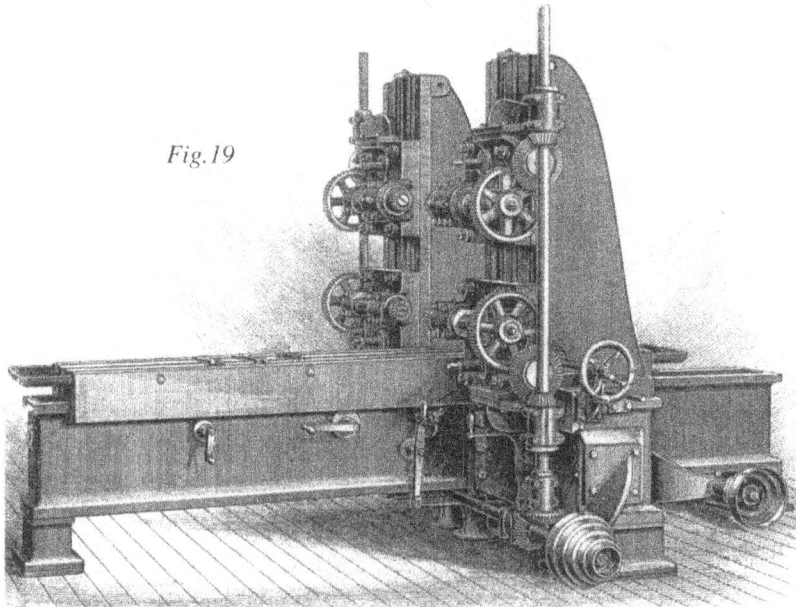

Fig.19

THE PRATT & WHITNEY FOUR-SPINDLE MILLING MACHINE.

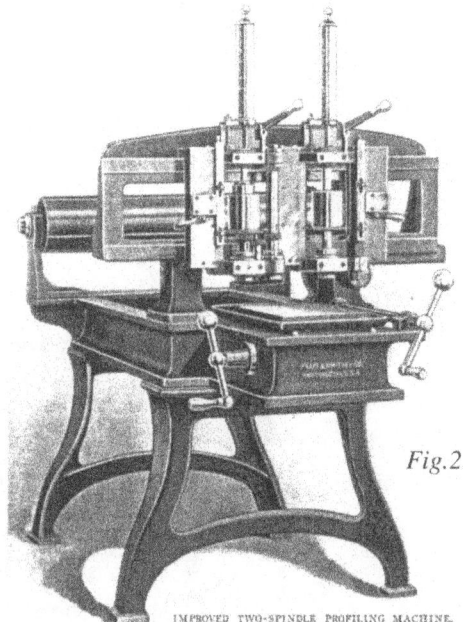

Fig.21

IMPROVED TWO-SPINDLE PROFILING MACHINE.

Improved column milling machines (Fig.20) were offered in 1898 in four sizes, and improved two-spindle profiling machines (Fig.21) in 1899 Milling machines offered in 1904, shortly after the Niles- Bement-Pond takeover, included one-spindle profiling machines (Fig.22) with 18" table travel and 19" cross travel; No. 2 die-sinking machines (Fig.23); No. 1 vertical spindle milling machines (Fig.24); No. 13 Lincoln millers (Fig.25) with 20" table travel; No. 2 Lincoln millers (Fig.26) with 12" table travel; No. 3 universal milling machines (Fig.27) with 36" table travel, 10" cross travel and 18½" vertical travel; No. 3 plain milling machines (Fig.28) with 36" table travel, 10 ½" cross travel and 20" vertical travel; No. 2 universal milling machines (Fig.29) with 29" table travel, 8" cross travel and 18½" vertical travel; No. 2 hand milling machines (Fig.30) with 5" table travel, 5" cross travel and 9" vertical travel; No. 1½ hand milling machines (Fig.31) with 3½" table travel, 5½" cross travel and 6" vertical travel; double column milling machines (Fig.32) with 28" x 7" tables; No. 3½ double horizontal milling machines (Fig.33); and thread milling machines in 6" x 14" size (Fig.34) and 6" x 80" size (Fig.35).

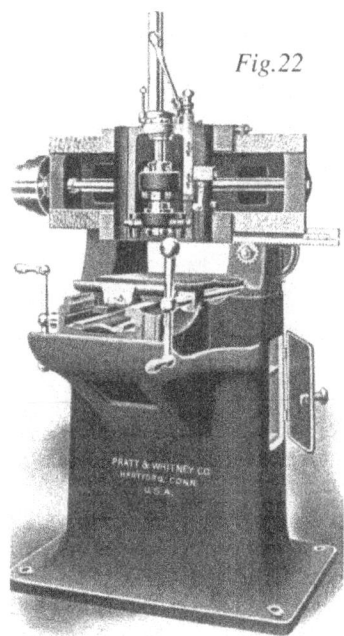

Fig.22

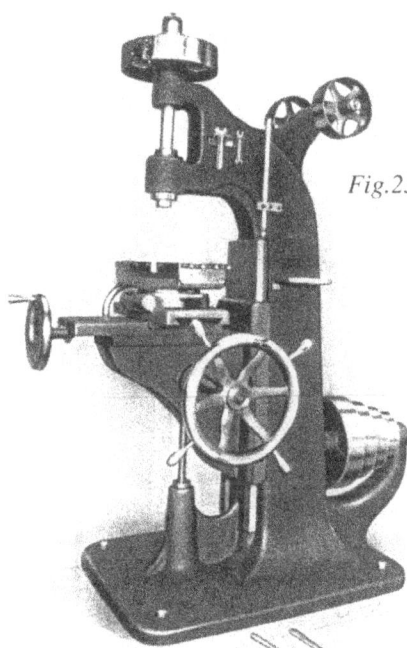

Fig.23

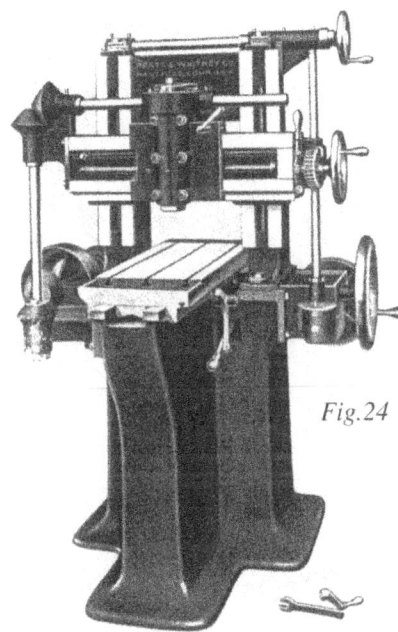

Fig.24

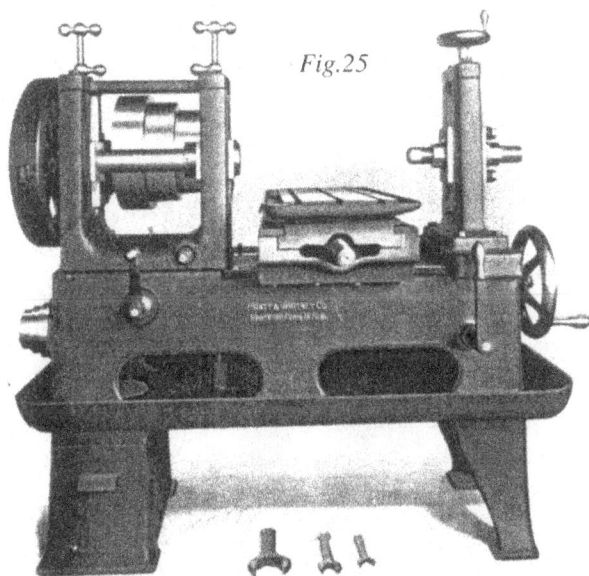

Fig.25

Fig.26

Fig.27

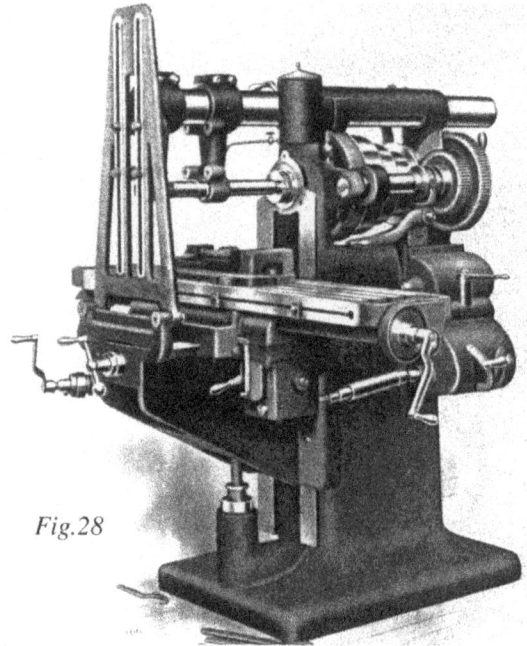

Fig.28

Fig.29

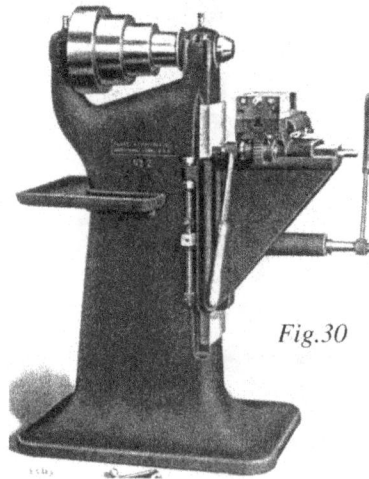

Fig.30

Fig.31

Fig.32

Fig.33

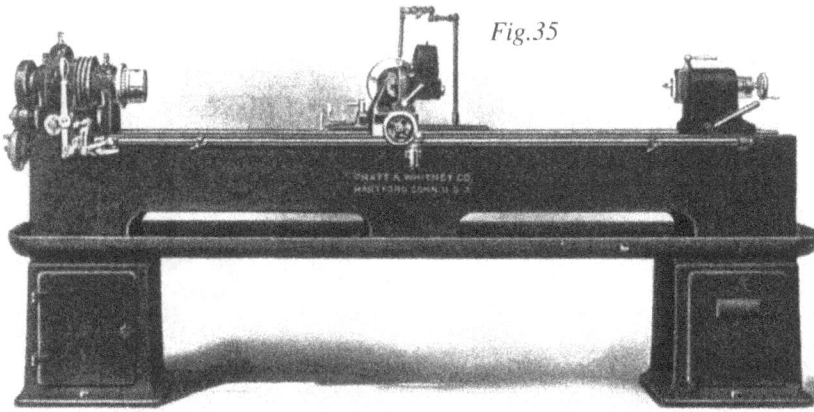

Fig.35

Fig.36

5-Inch Automatic Milling Machine   Belt Change Head
Patented March 24, 1916; June 30, 1916; October 16, 1916

Milling machine production continued through World War I with development concentrated on automatic machines (Fig.36) which allowed one man to operate multiple machines. Other milling machines offered in 1917 included No. 00 hand (Fig.37) with 14½" x 2" tables; No. 1 bench hand (Fig.38) with 14½" x 3½" tables; No. 10 hand (Fig.39) with 16" x 4¾" tables; and No. 2 column power (Fig.40) with 24" x 5" tables.

Fig.38

Fig.37

Fig.39

Fig.40

PREMIER MACHINERY CO., Milwaukee, WI

Builder, in 1915, of the keyseat miller *shown at right*. Quite small, the machine weighed only 275 pounds complete with the stand.

KEYSEAT MILLER

THE PROVIDENCE TOOL COMPANY'S MILLING MACHINE.

PROVIDENCE TOOL CO.,

Providence, RI

Formed in June, 1847, primarily to make nuts. Early in the Civil War, the firm was awarded contracts for military arms production. Frederick W. Howe, an expert machine designer who had worked at ROBBINS & LAWRENCE, was hired to oversee production and obtain the necessary machinery. In 1862, Howe approached Joseph R. Brown of J.R. BROWN & SHARPE with his need for a machine to mill twist drill flutes. The resulting machine, built by J.R. BROWN & SHARPE, was the first universal milling machine. By 1863, Howe had developed a production milling machine, *shown above*, which was built and sold by the Providence Tool Co. The Howe design milling machine became a product of J.R. BROWN & SHARPE by 1868, the same year Frederick W. Howe became a partner in the firm.

PUTNAM, J. & S.W., Fitchburg, MA

later **PUTNAM & CO., J. & S.W.,** Fitchburg, MA

later **PUTNAM MACHINE CO.,** Fitchburg, MA

A partnership of John Putnam (1810-1888) and his brother Salmon W. Putnam (1815-1872) formed in 1836 in Ashburnham, MA, moving to Fitchburg, MA, in 1838. Lathes were offered as early as 1838.

The brothers reorganized as J. & S.W. PUTNAM & CO. in 1850 when Charles H. Brown and Benjamin Snow were admitted as partners and again as the PUTNAM MACHINE CO. when John Q. Wright, Sylvester Wright and Charles Burleigh were admitted in August, 1854. Salmon W. Putnam served as president from 1858, when the firm became a stock company, until his death in 1872. John Putnam was not active in company management after 1858. Salmon W. Putnam, Jr. (1844-1923) managed the firm from 1872 until 1882 when, with his brother

George E. Putnam (1854-1892), he left to form a competing firm, the PUTNAM TOOL CO. (S.W. PUTNAM'S SONS).

The PUTNAM TOOL CO. was merged into the PUTNAM MACHINE CO. March 18, 1886, a few months after Salmon W. Putnam, Jr. and his brother rejoined and gained control of the PUTNAM MACHINE CO. Salmon W. Putnam, Jr. served as vice president until selling the firm to MANNING, MAXWELL & MOORE in 1913. George E. Putnam served as superintendent until his death in 1892.

The firm made a wide variety of machine tools including lathes, planers, shapers, and slotters. Milling machine production included column "slabbers" (Fig.1) offered in 1877 in two sizes, No. 0 and No. 1; and Lincoln type millers (Fig.2) also offered in 1877 and made in three sizes, No. 2, No. 3, and No. 4.

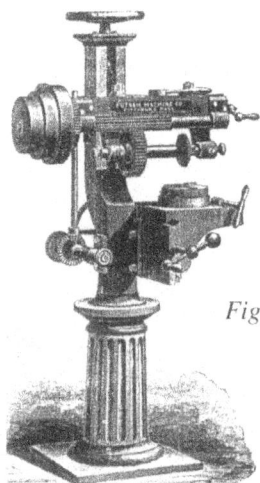

Fig.1

No. 1 Column Slabber.

Fig.2

No. 2 Milling Machine.

In 1892, the firm introduced a new design universal milling machine (Fig.3) which included the very unusual feature of a counterbalanced knee. Production, in slightly modified form (Fig.4), continued as late as 1901 and now included high speed milling attachments (Fig.5) with speeds to 3100 RPM, and vertical spindle milling attachments (Fig6). *(illustrations continued next page)*

Fig.3

A NEW UNIVERSAL MILLING MACHINE.

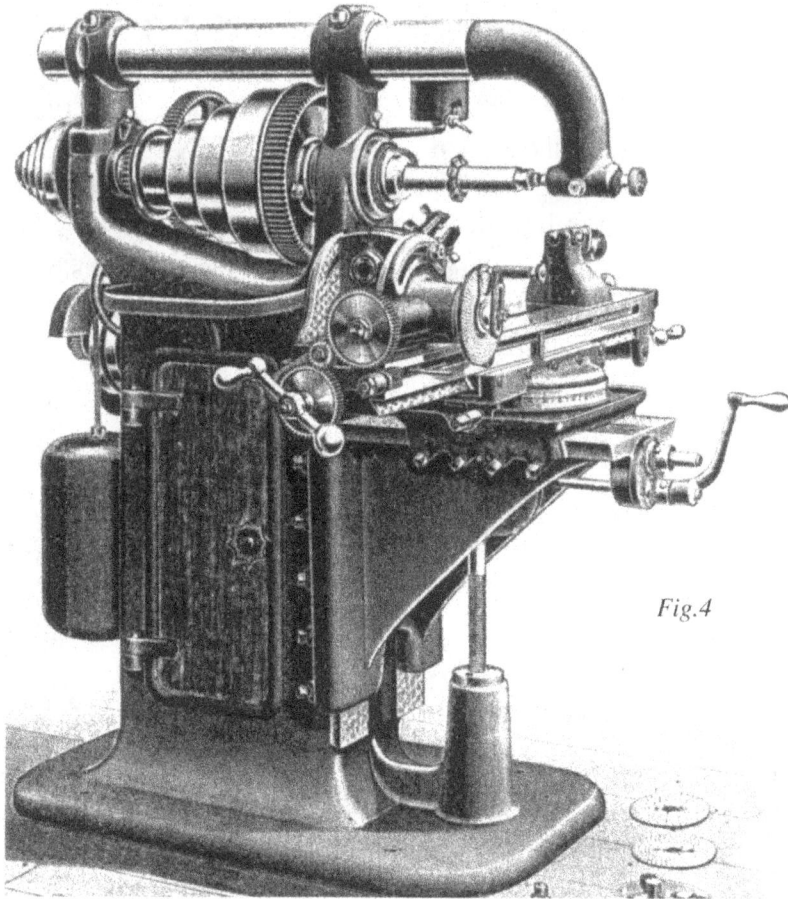

Fig.4

HIGH SPEED MILLING ATTACHMENT

Fig.5

VERTICAL SPINDLE MILLING ATTACHMENT

Fig.6

REED, F.E., Worcester, MA

later **REED & CO., F.E.,** Worcester, MA

later **REED CO., F.E.,** Worcester, MA

Formed in 1881 by Frederick E. Reed (1847-1917) as a reorganization of A.F. PRENTICE & CO. Reed had been a partner in A.F. Prentice & Co. before buying the firm in 1877. With an employment of 50 hands, Reed continued production of the Prentice machine tools.

The firm became F.E. REED & CO. when John R. Back was admitted as a partner January 1, 1891. It incorporated as the F.E. REED CO. in 1894 with 150 hands. Reed retired in 1912 when the firm was merged with Prentice Brothers Co. to form the REED-PRENTICE CO.

Milling machine production included an odd machine (Fig.1), probably first made by A.F. Prentice & Co. The machine was equipped with horizontal movement in the head stock, vertical, cross, and rotary movement in the workholding vise (also equipped with a dividing attachment), and a tail stock with set-over capability. It may have also been used as a taper cutting lathe.

By 1884, the No. 2 plain milling machine (Fig.2), which has features of the LINCOLN and ROOT milling machines, was offered. The machine weighed 2035 pounds and was priced at $450. A special, double head milling machine (Fig.3)for milling ends of structual iron was introduced in 1892 and a "New Pattern" knee type milling machine (Fig.4) in 1896.

Fig.1

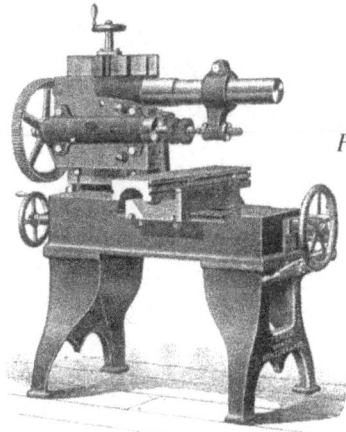

Fig.2

NEW MILLING MACHINES, BUILT BY F. E. REED, WORCESTER, MASS.

Reed's double-head milling-machine.

Fig.3

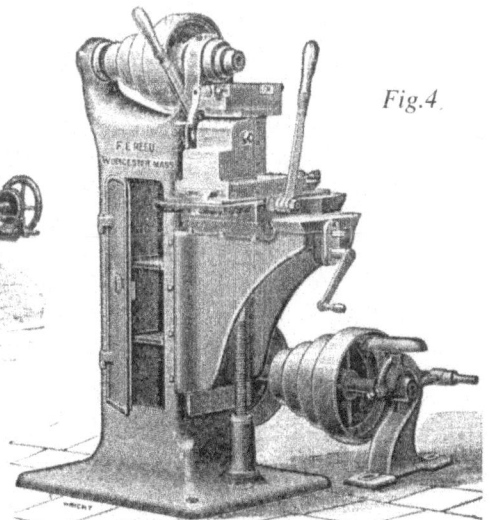

Fig.4

REED, JOHN C., Providence, RI

Maker, in 1864, of "a 28 inch index cutting engine just finished and for sale."

REED-PRENTICE CORP., Worcester, MA

Formed in 1912 by the merger of F.E. REED CO. and PRENTICE BROTHERS CO. Lathes and drilling machines formed the bulk of the new company's production. The BECKER MILLING MACHINE CO. and the Whitcomb-Blaisdell Machine Tool Co. were absorbed in 1923.

Milling machine production was revived in 1916 to meet demand from the greatly expanding armaments industry and included Lincoln millers (Fig.1), external thread milling machine (Fig.2) and internal thread milling machines (Fig.3).

  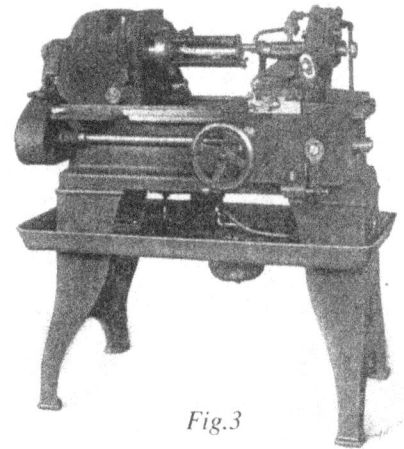

Fig.1   *Fig.2*   *Fig.3*

Lincoln Miller with Automatic Feed and Table Return   External Thread Milling Machine for threading Rifle Barrels   Internal Thread Milling Machine for threading Rifle Receivers

RESEK MACHINE TOOL CO.,
Cleveland, OH

Founded in 1903, the firm probably failed in 1905. Products included a Duplex lathe and an automatic valve milling machine as *shown at right*.

RICE & ANDERSON, Quincy, IL

Builder, in 1871, of a stone dressing machine *(below)*, patented July 12, 1870, by A.G. Anderson. Although made for cutting stone rather than metal, the machine is clearly a milling machine and was equipped with inserted tooth cutters much like the Ingersoll cutters patented i9 years later.

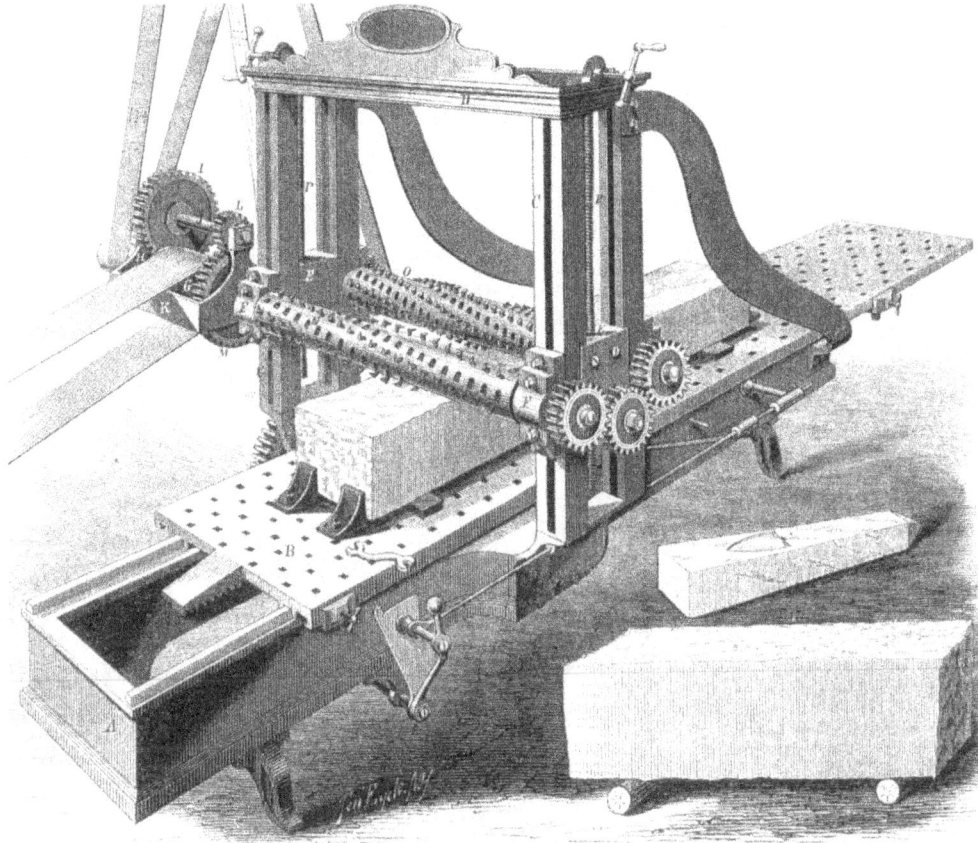

ANDERSON'S STONE DRESSING MACHINE.

RICHARDS MFG. CO., Brooklyn, NY

Maker, in 1895, of small bench milling machines *(right)*. The machine was only 12½" high and weighed 30 pounds.

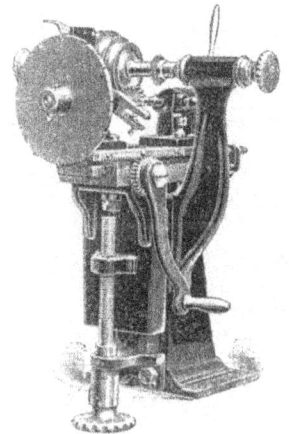

BENCH MILLER.

RIEHLE BROTHERS TESTING MACHINE CO., Philadelphia, PA

A major builder of testing machines for metallurgical work, the firm offered, in 1897, a special milling machine for preparing test pieces for the testing machines. Both sides of the piece were machined in the same pass, using two 6" cutters rotating in opposite directions. *(illustration on next page)*

Riehle

RIVETT LATHE MFG. CO., Boston, MA

later RIVETT LATHE & GRINDER MFG. CO., Boston, MA

Formed about 1900 as a reorganization of the FANEUIL WATCH TOOL CO. operated by Edward Rivett (1851-1937). Rivett sold out in 1912 and the firm reorganized as the Rivett Lathe & Grinder Mfg. Co.

Precision bench lathes were the primary product, but the firm also offered bench millers with swivel tables (Fig.1), table and cross travels of 6", and weighing 250 pounds. The machine was also offered with a back geared head (Fig.2).

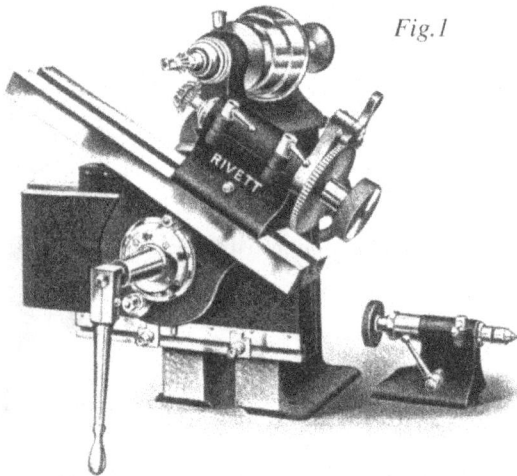

Fig.1

Rivett Bench Miller with Swivel Table for Angles: weight, 225 lbs.

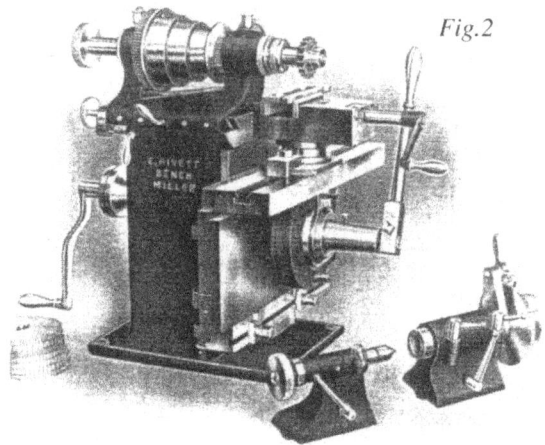

Fig.2

Rivett Swivel Bench Miller with Back Geared Head

ROBBINS & LAWRENCE CO., Windsor, VT

Formed in 1850 by Samuel E. Robbins (1810-1874) and Richard S. Lawrence (1817-1892) as a reorganization of Robbins, Kendall & Lawrence. Primary products were a variety of firearms, most made on contract for the U.S. and British governments. Machine tools soon became a secondary product when they began selling machines which had been designed for their own firearms production.

Robbins, who was a Bangor, ME, businessman, served as president and Lawrence as superintendent until the firm failed in 1856. The plant and equipment were bought by Ebenezer G. Lamson in 1859 who then organized LAMSON, GOODNOW & YALE. Lawrence moved to Hartford, CT, where he became superintendent of the Sharps Rifle Mfg. Co.

Early machine tool production included plain milling machines (Figs.1-2) and edging or profiling machines (Fig.3) both designed in 1848 by Frederick W. Howe (1822-1891). Later production included index milling machines designed by Howe about 1850 (Fig.4), some of which were still in use at the Springfield Armory in 1900 (Fig.5) and later.

The plain milling machine evolved into the ROOT milling machine made by the COLT PATENT FIREARMS MFG. CO. and the LINCOLN milling machine developed at the PHOENIX IRON WORKS.

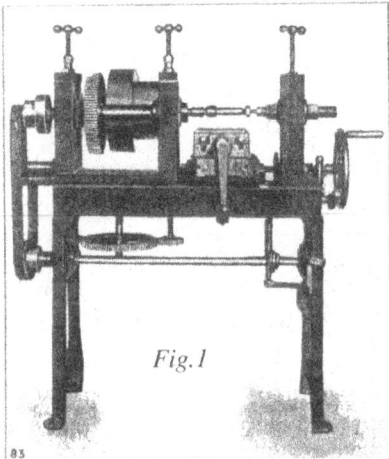

Fig.1

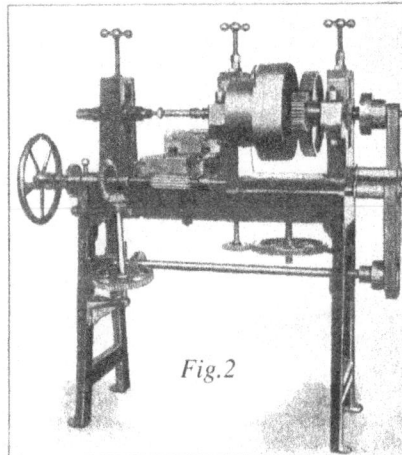

Fig.2

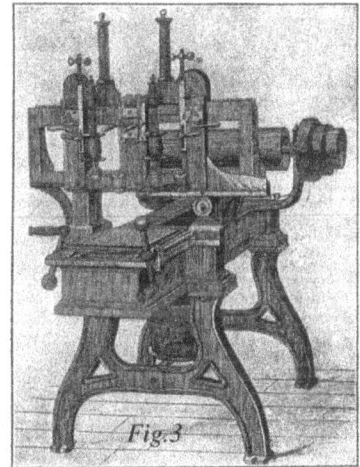

Fig.3

Fig.4

Fig.5

ROBBINS, L., Worcester, MA

later **ROBBINS MACHINE CO.,** Worcester, MA

Founded by Lafayette Robbins about 1893 to make lathes and other machine tools. In 1908, he incorporated as the ROBBINS MACHINE CO. which operated into the 1920s.

Milling machine production included a profiling and milling machine offered in 1920. As *shown at right*, it was equipped with 12" table travel, 7" cross travel and 6" knee travel coupled with $2\frac{3}{4}$" spindle travel.

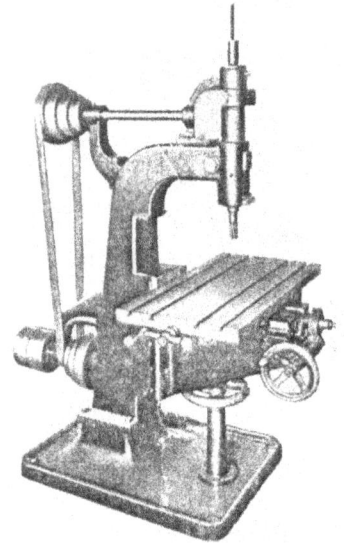

ROBERTSON, WILLIAM H.   See Whitney Armory

ROCKFORD MACHINE TOOL CO., Rockford, IL

Formed in 1905 as a reorganization of the National Engine Co. Robert C. Lind (1856-1935) was president until his death in 1935.

The firm's primary products were shapers and planers, but in 1911 a heavy vertical milling machine, *shown at left* was introduced. Weighing 6800 pounds, the machine had 48" table travel, $14\frac{1}{2}$" cross travel and 21" vertical travel via a sliding head.

By 1919 the milling machine was a product of WM. K. STAMETS.

ROCKFORD MILLING MACHINE CO., Rockford, IL

Formed in 1910 as a reorganization of the NATIONAL MACHINE & TOOL WORKS. Oscar J. Sundstrand (1889-1972) and his brother-in-law Edwin Cedarleaf operated the company, joined in 1912 by Oscar's brother, Gustaf D. Sundstrand (1880-1930). In 1926, the firm merged with the Rockford Tool Co. to form the Sundstrand Machine Tool Co.

First production was a continuation of the hand millers built by the National Machine & Tool Works. A new, heavier machine (Fig.1) with a vertical milling attachment, 24" table travel, $7\frac{1}{2}$" cross travel and 19" vertical travel was introduced in 1911. A slotting attachment (Fig.2) was also offered.

In 1913, a No. 2 plain with single pulley drive (Fig.3) was introduced. The machine was offered in 1914 as a duplex miller (Fig.4) by adding a second spindle head driven by shafting running through the overarm. The duplex appears not to have been a commercial success. *(continued on next two pages)*

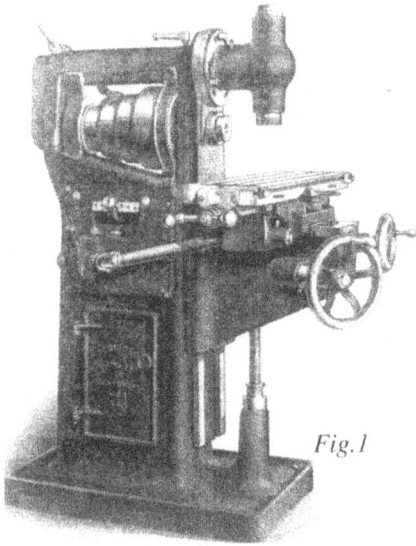

Fig.1

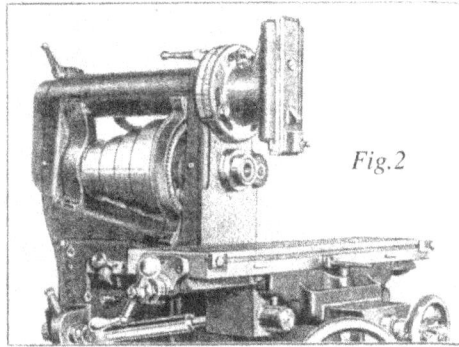

Fig.2

Slotting Attachment applied to Milling Machine

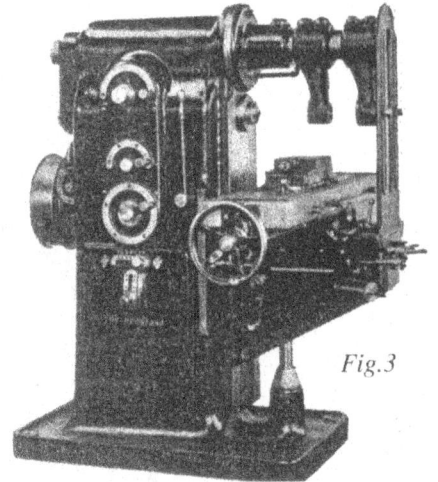

Fig.3

Products offered in a 1915 catalog included; No. 1 hand millers (Fig.5), No. 2 hand millers with overarm (Fig.6), No. 3 hand millers with flange mounted overarm (Fig.7), No. 0 plain milling machines (Fig.8), No. 1 plain milling machines (Fig.9), No. 2 universal milling machines (Fig.10), and No. 2 heavy duty double back geared milling machines (Fig.11).

A new design No. 3 plain milling machine (Fig.12) was introduced in 1920. The most obvious new feature was a rectangular overarm, unique to this design.

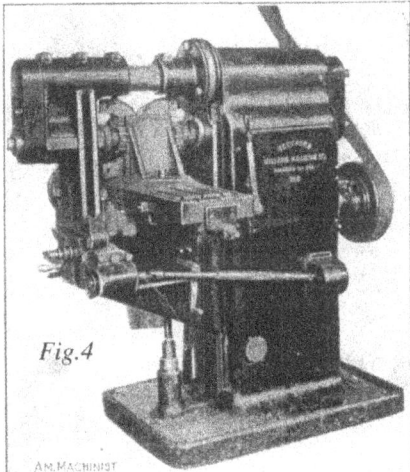

Fig.4

AM.MACHINIST

No. 1. FOR BENCH.

Adjustments 10x4x6¼ Inches.

No. 2. WITH OVERHANGING ARM.

Adjustment 10x4x6¼ Inches.

Fig.5

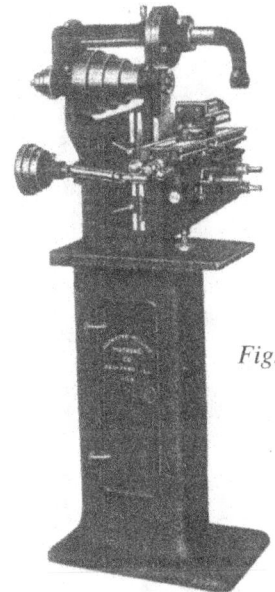

Fig.6

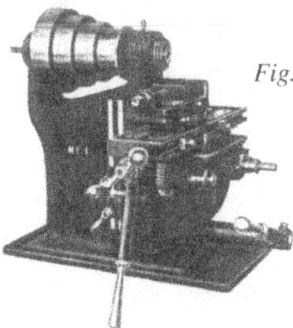

Fig.7

Fig.8

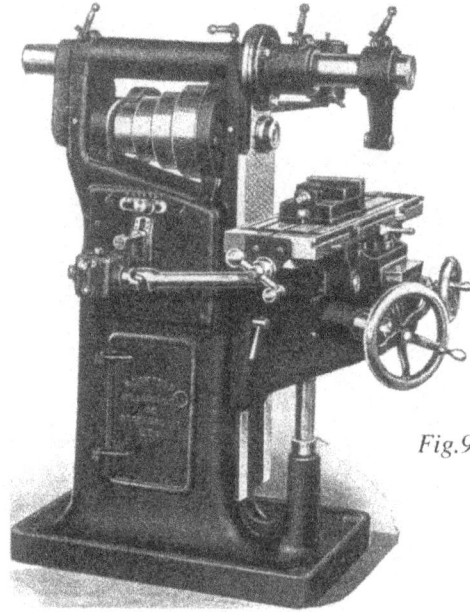

Fig.9

HEAVY DUTY PLAIN DOUBLE BACK GEARED CONE DRIVEN.

Fig.11

Fig.10

Fig.12

ROWBOTTOM MACHINE CO., Waterbury, CT

Founded in 1902 by George Rowbottom (1867-1953) to make grinding machines and light machinery. He is credited with developing the first practical cam cutting machine about 1915.

By 1916, Rowbottom had introduced the universal cam milling machine, *shown at right*, which could cut box or face cams and drum or side cams up to 28" diameter.

Production continued for many years as the company continued to operate into the 1960s.

RYERSON-CONRADSON

See Conradson Machine Tool Co.

SELLERS & CO., WILLIAM, Philadelphia, PA

A partnership of William Sellers (1824-1905) and his brother John Sellers, Jr. (1826-1906), formed in 1855 when Edward Bancroft died and the partnership of Bancroft & Sellers was reorganized. Their cousin, Coleman Sellers, who had served as superintendent at NILES & CO., joined the firm in 1856 as chief engineer. William Sellers was elected president when the firm incorporated in 1886, serving until his death in 1905.

Production centered around large machine tools for the railroad industry. Only one type of milling machine was built; a small manufacturing miller with 14" table travel, 4½" cross travel and 4" vertical travel, as *shown at left*, offered from about 1870 to 1880.

SILVER & GAY, North Chelmsford, MA

later **SILVER & GAY CO.,** North Chelmsford, MA

A partnership of Harvey Silver (?-1884) and Ziba Gay, Jr. (1823-1902) formed in 1857 as a reorganization of GAY, SILVER & CO. In 1884, the firm reorganized as the SILVER & GAY CO.

Products included textile machinery, milling machines, lathes, planers and other machine tools. Machine tool production ceased before 1901.

SLATE, DWIGHT, Hartford, CT

later **SLATE MACHINE CO., DWIGHT**, Hartford, CT

Formed by Dwight Slate (1816-1906) in 1876. In 1886 Slate bought the Ballou Mfg. Co. and reorganized as the Dwight Slate Machine Co. where he served as president until his death in 1906.

Production centered around drilling machines, but Slate also offered, beginning in 1884, small plain milling machines (Fig.1) with 20" table travel, priced at $250; and nut and bolt head millers (Fig.2) priced at $180 complete with chuck. Slate also offered a No. 2 automatic gear cutter (Fig.3) based on the milling machine design. All the above machines were offered at least as late as 1904.

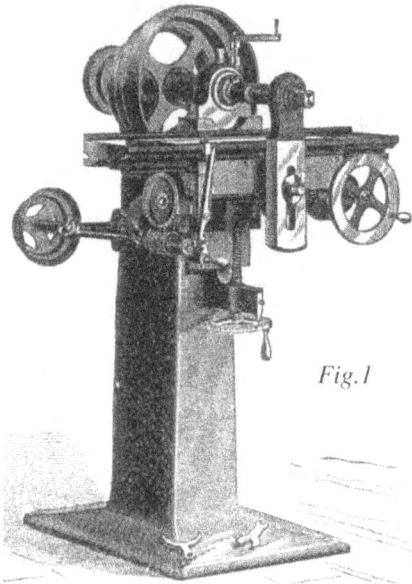

Fig.1

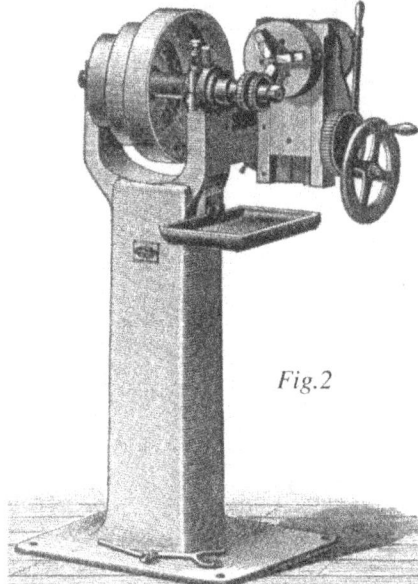

Fig.2

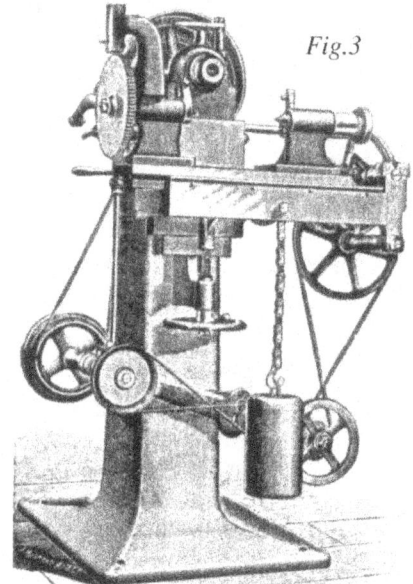

Fig.3

SLOAN & CO., C.T., Newark, NJ

later **SLOAN, CHACE & CO.**, Newark, NJ

later **SLOAN & CHACE MFG. CO.**, Newark, NJ

Founded by Charles T. Sloan about 1875. In 1886 the firm reorganized as Sloan, Chace & Co., a partnership of Charles T. Sloan and George E.O. Chace. By 1906, the firm became the Sloan & Chace Mfg. Co. All three firms made small bench lathes; the latter two also made bench milling machines with 7" vertical travel, 10" horizontal and 4¾" cross travel.

Sloan & Chace Bench Milling Machine for Precision Work.

SMALLEY-GENERAL CO., Bay City, MI

Builder, beginning in 1916, of thread milling machines developed for the World War I munitions industry. Thread milling machine production continued until 1930, when the company was absorbed by the Wickes Brothers Co.

The machine *shown on the next page* was equipped with a collet chuck for threading the nose of 8" shells; other chucks could be furnished for other shell sizes. *(see next page)*

Smalley-General

Thread Milling Machine built by the Smalley-General Co.

SMITH & GARVIN, New York, NY

Founded in 1862 by Hugh R. Garvin. The firm reorganized in 1879 as E.E. GARVIN & CO.

Products in 1873 included drill presses, milling machines and small lathes.

SNOW, BROOKS & CO., Meriden, CT

Partnership of Gamaliel F. Snow, James S. Brooks and Charles Parker, formed in 1856 as a reorganization of the MERIDEN MACHINE CO. The firm continued production of milling machines previously made by the Meriden Machine Co. and, prior to that, by OLIVER SNOW & CO.

By 1862, the firm had reorganized as PARKER, SNOW, BROOKS & CO.

SNOW & CO., OLIVER, Meriden, CT

Formed about 1848 to make hand planing machines. By 1852 Snow was advertising milling machines, larger planers and hand lathes.

In 1854, the firm was reorganized as the MERIDEN MACHINE CO.

STAMETS CO., WM. K., Pittsburgh, PA

Founded in 1912 by William K. Stamets (1884-1958) as a machinery sales company.

By 1918, Stamets began building heavy machine tools, or more likely had them built for him by others. These machines included the No. 44 heavy vertical milling machine, *shown at right*. The machine is clearly identical to one introduced by the ROCKFORD MACHINE TOOL CO. in 1911.

STANDARD ENGINEERING WORKS, Pawtucket, RI

Formed in 1916 to make hand milling machines. The first model (Fig. 1 had table travel of 14" and cross travel of 6". A combination hand or power feed model (Fig.2) of the same size was introduced in 1917. Production probably ceased with the end of World War I.

Fig.1

HAND MILLER WITH WEIGHT FEED

Fig.2

POWER- OR HAND-FEED MILLING MACHINE

STARK TOOL CO., Waltham, MA

Founded May 1, 1862, by John Stark to make small lathes and other machinery for the watch industry. Stark served as president until his death about 1900. His son John Stark, Jr. then took over the firm, operating as proprietor until incorporating in 1918.

Early production centered around watchmakers' lathes patented May 30, 1865, reissued October 19, 1872.

Small milling machines offered in its 1902 catalog included a profiling machine (Fig.1); plain milling machine (Fig.2) priced at $250, available with a dividing head priced at $75; and a universal milling machine, pinion and wheel cutter combined (Fig.3) priced at $425.

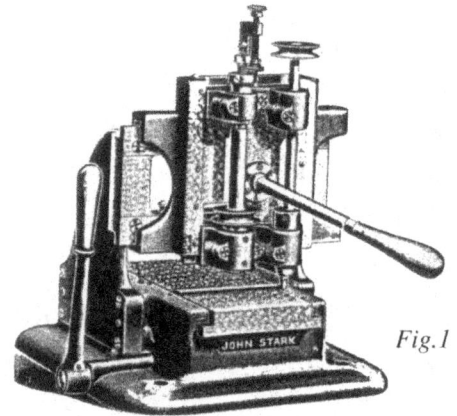

Fig.1

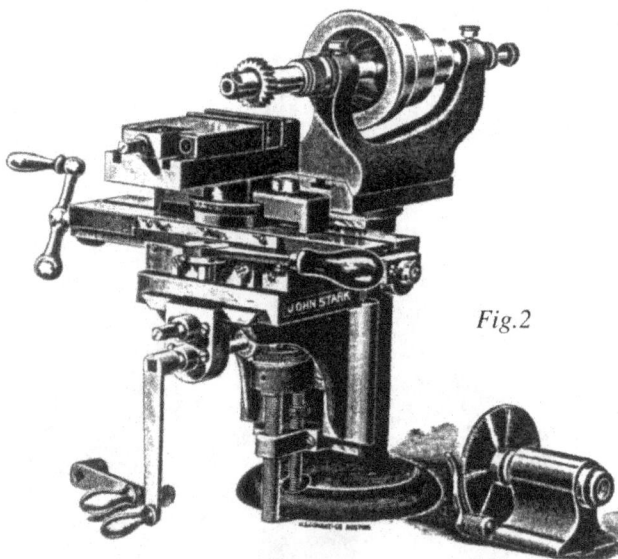

Fig.2

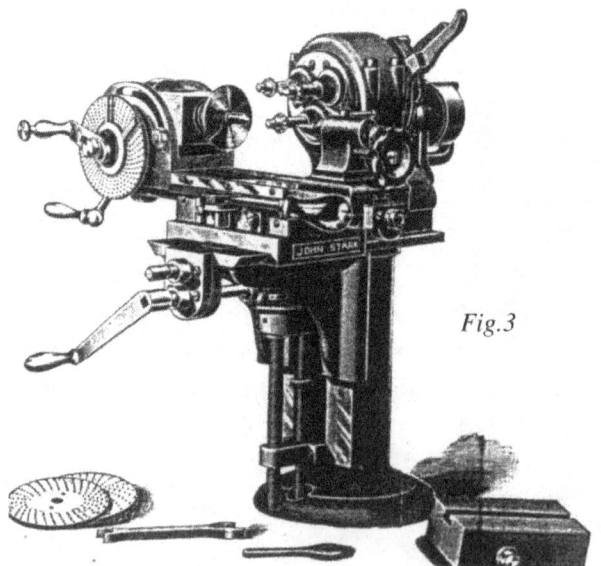

Fig.3

STATES MACHINE CO., Newark, NJ, later Hartford, CT

Formed about 1890 by Abraham Vanderbeek to make milling machines, vises and universal joints. 1890 production included a combination boring and milling machine (Figs.1-4), equipped with hand or power feed for the spindle when boring, and power feed for the table when milling. A combination dividing head and angle table (Fig.5) was offered. A circular milling attachment (Fig.6) was introduced in 1891.

Fig. 1.—Boring Side.

Fig. 2.—Milling Side.

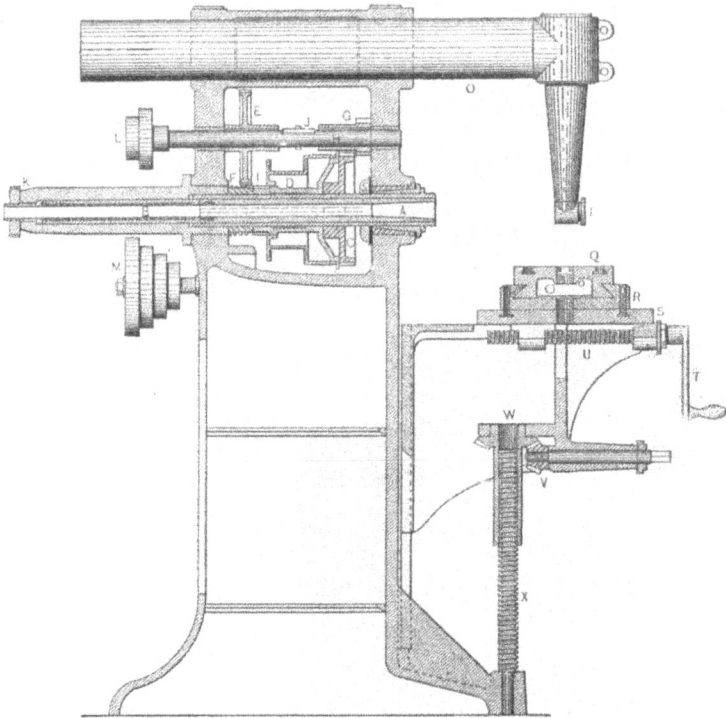
Fig. 3.—Vertical Section through Spindle.

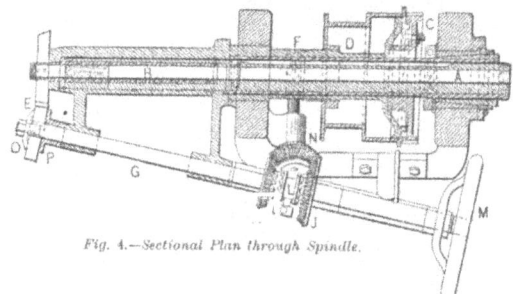
Fig. 4.—Sectional Plan through Spindle.

Fig. 5.—Dividing Head and Angle Table.

A new milling machine (Fig.7), now without the boring function and therefore with much improved spindle stiffness, was introduced in 1892. Both the circular milling attachment (Fig.8) and the combination dividing head and work table (Fig.9) were offered with the new machine. The firm reorganized as the VANDERBEEK TOOL WORKS in 1898.

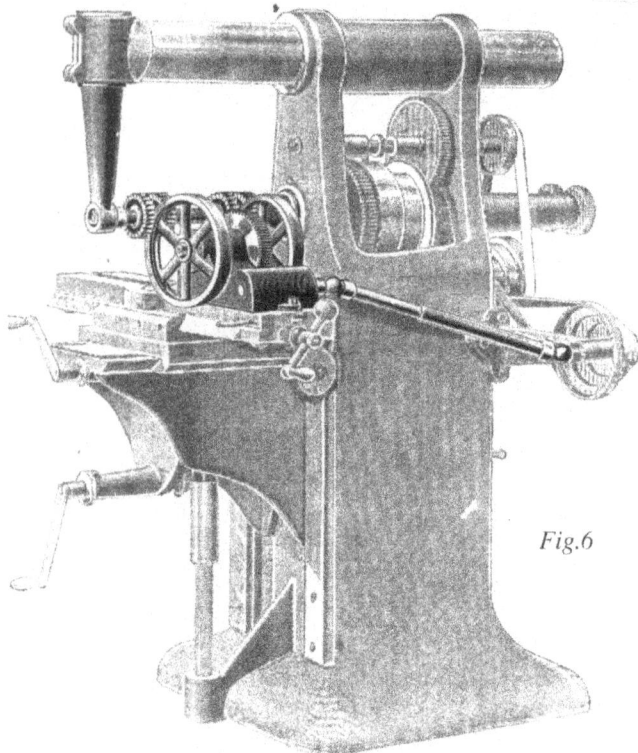

Fig.6

CIRCULAR MILLING ATTACHMENT.

Fig.7

NEW MILLING MACHINE.

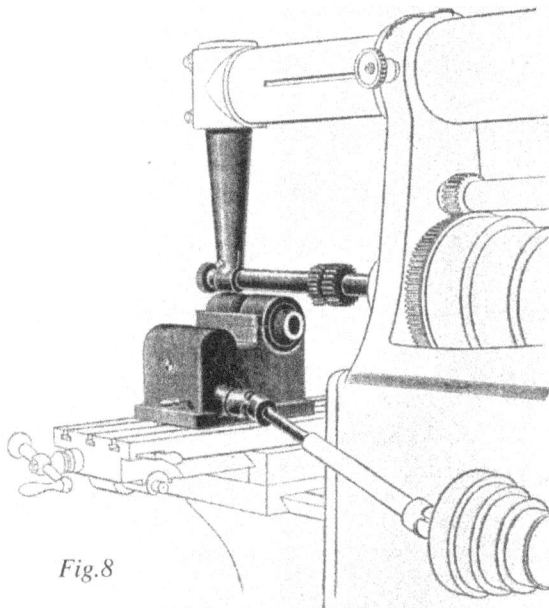

Fig.8

COMBINED DIVIDING HEAD AND WORK TABLE.

Fig.9

STEPTOE CO., JOHN, Cincinnati, OH

Formed in 1913 as a reorganization of the Steptoe Shaper Co.

The new firm continued production of the Steptoe shaper line and introduced a new line of hand milling machines as *shown at right.*

Operations continued until 1927 when the firm was absorbed by the Western Machine Tool Works.

IMPROVED HAND MILLER

SUPERIOR MACHINE & ENGINEERING CO., Detroit, MI

Builder, beginning in 1916, of hand milling machines (Fig.1) weighing 750 pounds and oscillating-head millers (Fig.2) with table working surface of only 3½" x 5". The spindle had an oscillating motion rather than steady rotation. Superior claimed "the machine is especially suitable for milling small slots, as the oscillating movement of the cutter prevents chips from jamming between the cutter and the work."

Fig.1

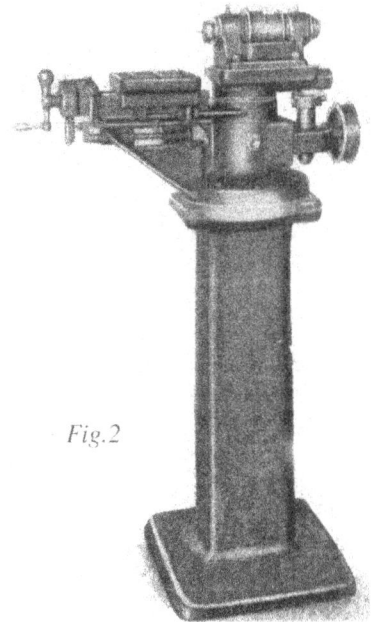

Fig.2

SWEETLAND & CO., New Haven, CT

Builder, in 1883, of the GRISWOLD bench milling machine *(right)*. The built-in vise could be both rotated and tilted for cutting gears or other index work.

BENCH MILLING MACHINE.

TAYLOR & FENN CO., Hartford, CT

Formed in 1907 by merger of Fenn Machine Co. and the PHOENIX IRON WORKS. Charles L. Taylor was treasurer, W.L. Fenn works manager. Light drilling machines were the primary product.

A light vertical milling machine, *shown right*, was introduced in 1917. The machine was equipped with a 4000 RPM spindle and 11" table travel, 6" cross travel, and 9 1/2" vertical travel.

THURSTON MFG. CO., Providence, RI

Operated by Horace Thurston (1850-1922), the company was formed in 1886 to succeed Thurston, Stearns & Co.

The firm specialized in machinery for jewelry makers, including an inverted milling machine, patented July 22, 1890, and designed for milling dies. The machine was made, in slightly modified form, as late as 1918.

VERTICAL MILLING MACHINE

Thurston Inverted Milling Machine for Milling Dies, used on Gun work, Sewing Machines, Jewelry, Drop Forgings; in fact, any work requiring Punching or Trimming Dies. Also, Contractors for Light Machinery and Special Tools.

Circular sent on application.
Correspondence solicited. . .

THURSTON MFG. CO.,
419 Eddy St., PROVIDENCE, R. I., U.S.A.

PAT. JULY 22, 90.

TOLEDO MILLING MACHINE CO., Toledo, OH

Formed in 1919 to make a No. 3 vertical milling machine with 42" table travel, 14" cross travel, and 14" vertical travel with another $6\frac{1}{2}$" in the spindle *(right)*. The firm announced that this was the first of a line to be developed, but no other models appear to have been introduced prior the failure of the company in 1926.

TRASK MILLING MACHINE CO., Lynn, MA

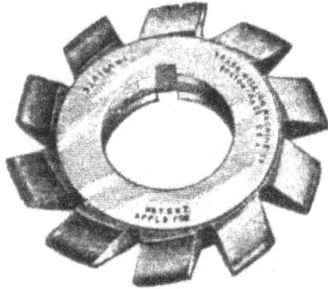

Formed about 1888 by C.H. Trask who received milling machine patents on March 6, 1888, October 29, 1889, and March 4, 1890. By 1892, the firm's ads referred only to milling cutters.

UNION VISE CO., Boston, MA

Formed in 1865 to make vises and milling machines (Fig.1) designed by Amos H. Brainard (1824-1905). By 1871, the firm was advertising a variety of milling machines (Fig.2).

The company was burned out in 1871; Brainard then reorganized as the BRAINARD MILLING MACHINE CO., which continued production of the same line of milling machines.

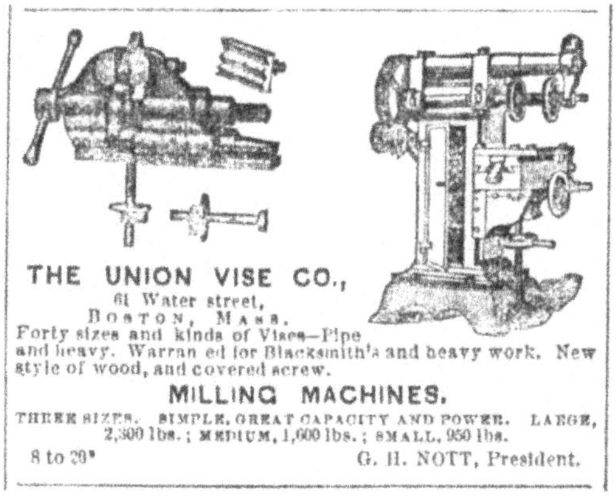

Fig.1

Milling Machine,

INDEX, STANDARD, UNIVERSAL, AND HORIZONTAL.—The largest variety to be found in the country, on hand and finishing. Workmanship, Material, and Design unsurpassed. Machines on exhibition at Fair of American Institute. UNION VISE CO. OF BOSTON. Office 80 Milk st. Works at Hyde Park, Mass.

Fig.2

UNITED STATES MACHINE TOOL CO., Cincinnati, OH

Organized in 1916, the firm took over the plant and business of the AMERICAN MILLING MACHINE CO. in 1917. Production seems to have been limited to its No. 1 hand miller (Fig.1) and vertical milling attachments (Fig.2).

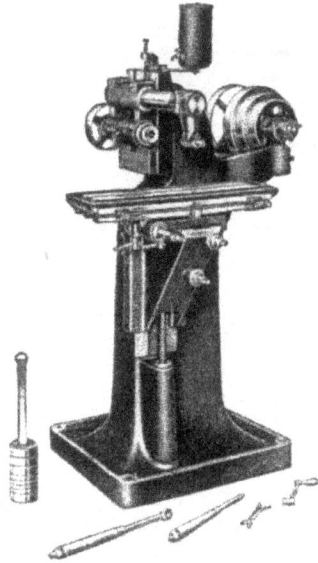

U. S. IMPROVED HAND MILLER

There are many milling operations done on large expensive milling machines that can be done in much less time on a Hand Milling Machine and at considerable less cost. Much more work can be done and more economically, on small parts. The quick adaptability of the machine, and the wide range of work, makes it useful.

Two Speed Friction Countershaft can be furnished if desired at extra price.

We are prepared to furnish all accessories, such as collets of all kinds, upright and flat vises, arbors.

Machine can also be arranged with motor drive and oil pump attachment.

Fig. 534A.

Fig.1

Description

No. 9 Brown & Sharpe Taper Hole in spindle regularly furnished.
Working Surface of the Table, $6\frac{5}{8}$x23 inches over all.
Adjustment of Table: Under Spindle, $15\frac{3}{4}$ inches;
                   In line with Spindle, 5 inches.
Table Feed with Hand Lever, 6 inches; with crank, 10 inches.
Top of Table can be brought on level with center of Spindle.
Top of Table furnished with standard $\frac{5}{8}$-inch T Slot.
Adjustable lever feed on spindle head, 4 inches.
Can be furnished with either three or two-step cone.
Net Weight . . . . 1000 lbs.

No. 1 PLAIN MILLING ATTACHMENT IN MACHINE, ALSO FLAT MILLING MACHINE VISE.

Fig.2

The spindles of these attachments are of crucible steel, hardened and ground with No. 7 B. & S. Taper hole in lower end clear through for drawbolt. Bearings are of special bronze and carefully scraped to fit. Adequate provision is made for compensation for wear.

Can also furnish No. 2 Universal Milling Attachment, if desired.

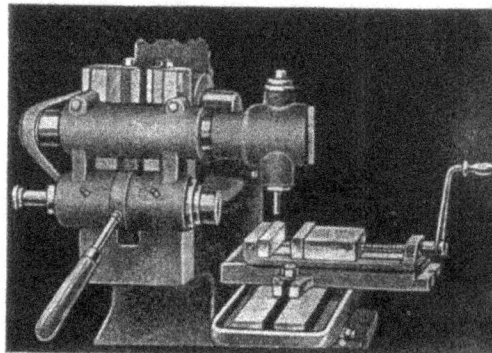

VANDERBEEK TOOL WORKS, Hartford, CT

Formed in 1898 as a reorganization of the STATES MACHINE CO. The firm continued production of the VANDERBEEK milling machine previously made by the States Machine Co. in addition to turret drills, vises and universal joints. The firm was active as late as 1917. *(see illustration next page)*

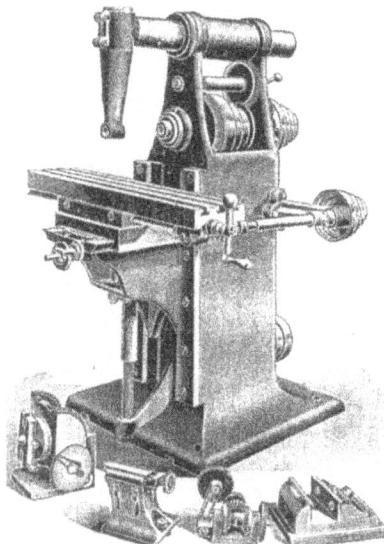

Vanderbeek

Tool Room and Shop
Milling Machines

Designed for both axial and radial milling, and with special reference to work production, stability and wear. Many features in our construction, not ordinary, will be found valuable.

....

A small line for immediate delivery.
....

Vanderbeek Tool Works
Hartford, Conn.

Manning, Maxwell & Moore, New York.

VAN NORMAN MACHINE TOOL CO., Springfield, MA

Formed in 1912 as a reorganization of the WALTHAM WATCH TOOL CO. The company was operated by brothers Charles E. Van Norman (1859-1946), Frederick D. Van Norman (1862-1963), and Edmund F. Van Norman.

Van Norman continued production of the bench lathes, milling attachments (Fig.1) with vises (Fig.2) and index heads (Fig.3), and DUPLEX milling machines made by the Waltham Watch Tool Co. The DUPLEX milling machines remained basically unchanged until the mid-1920s when new models were introduced.

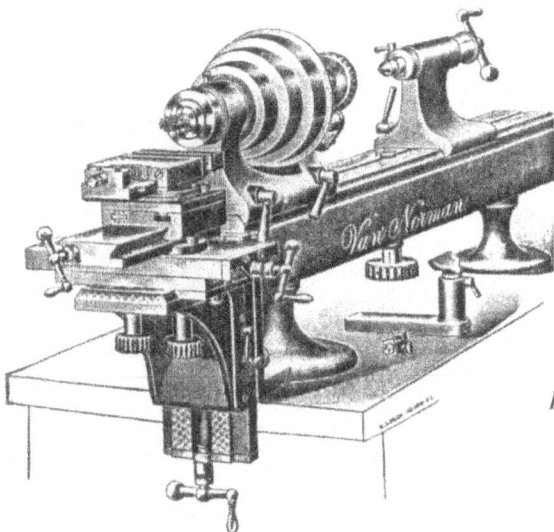

Fig.1

Vise   Fig.2

Fig.3

Index Head

VERTICAL DIE SHAPER CO., Rochester, NY

Maker, in 1901, of an attachment *(on right)* for converting a milling machine to a die shaper with a $1\frac{3}{8}$" stroke.

WAINWRIGHT & KELLY, Trenton, NJ

Builder, in 1907, of a small, plain milling machine as *shown at left*. The machine had table travel of 18", cross travel of $4\frac{1}{2}$" and vertical travel of 13"; weight was 1100 pounds.

WALTHAM MACHINE WORKS, Waltham, MA

Formed in 1898 as a partnership of Edmund L. Sanderson and Benjamin F. Ellis to make small lathes and other small machine tools for the Waltham watch industry. The firm expanded in 1902 by taking over the factory of the ALCORN & AMES CO. Products included thread millers offered from about 1900 to the 1920s. *(see right)*

WALTHAM WATCH TOOL CO., Waltham, MA, later Springfield, MA

Formed in 1885 as a reorganization of the Hopkins Watch Tool Co. The organizers were Caleb Hopkins Van Norman, founder of the Hopkins Watch Tool Co. and his sons Edmund F. Van Norman (?-1890) and Frederick D. Van Norman (1862-1963). A third son, Charles E. Van Norman (1859-1946) joined in 1888. The firm relocated to Springfield, MA, in 1890. *(continued next two pages)*

Early products were aimed at watchmakers and included bench lathes with milling attachments and may have included bench milling machines. Production of larger milling machines began in 1896 with the introduction of the DUPLEX milling machine, patented March 9, 1897, and August 17, 1897, by Charles E. Van Norman. The most important feature of the new design was a swiveling spindle head which could be operated in a horizontal position, vertical position, or any point between.

No. 0 machines (Fig.1) with 19" table travel, 5" cross travel and 12' vertical travel; and No. 2 machines (Figs.2-3) with 28" table travel, 10" cross travel and 18" vertical travel were the first models introduced. A No. 1 machine (Figs.4-6) with 22" table travel, 8½" cross travel and 14" vertical travel was introduced in 1901. Note the complex rope drive which was soon dropped. Profiling attachments (Fig.7) and index centers (Fig.8) were offered for all machines.

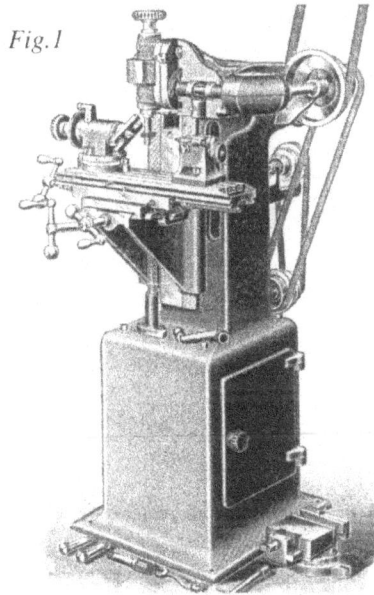
Fig.1

Fig.2

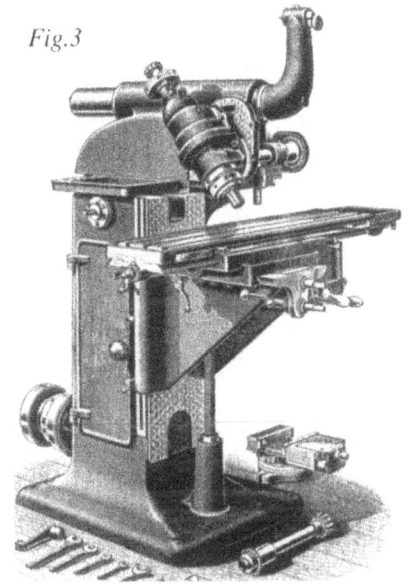
Fig.3

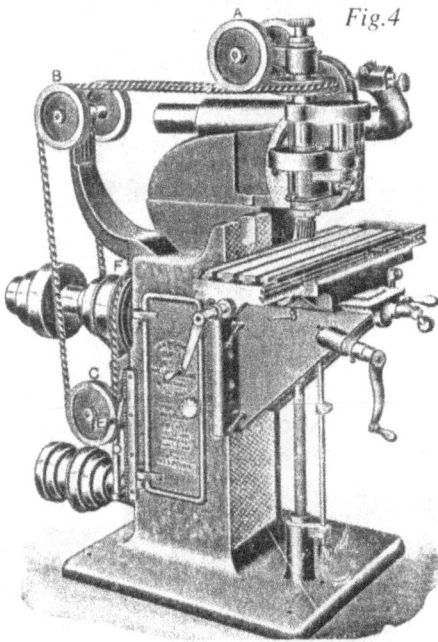
Fig.4

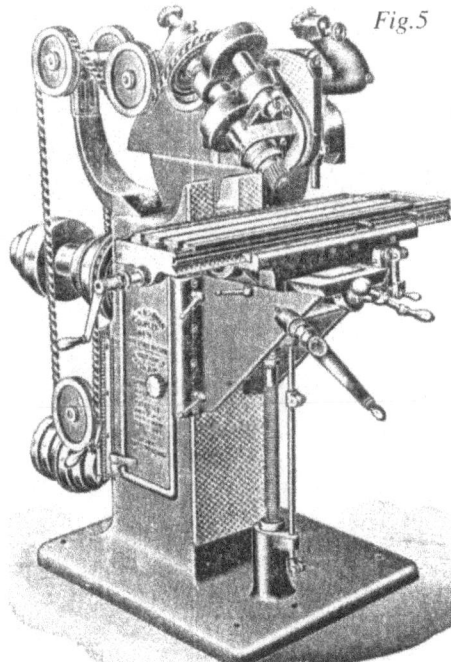
Fig.5

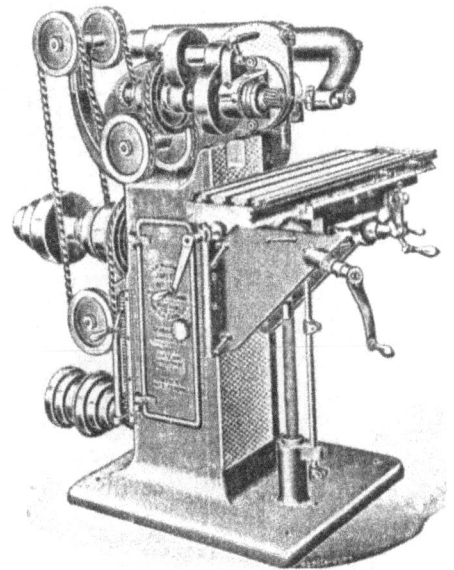
Fig.6

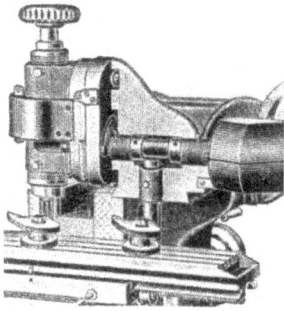

Fig.7

Fig.8

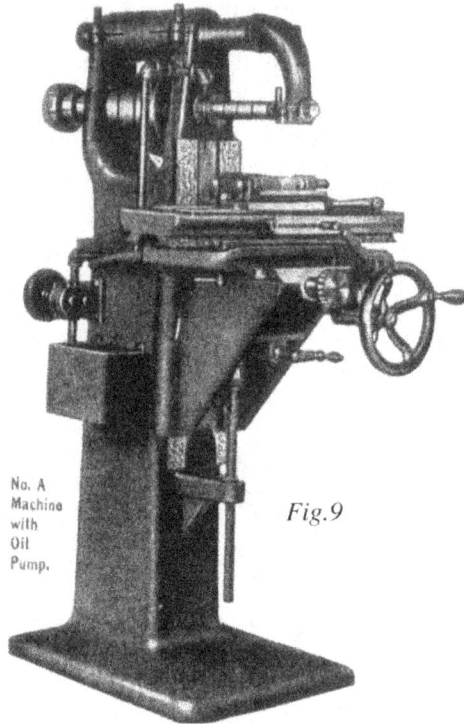

No. A Machine with Oil Pump.

Fig.9

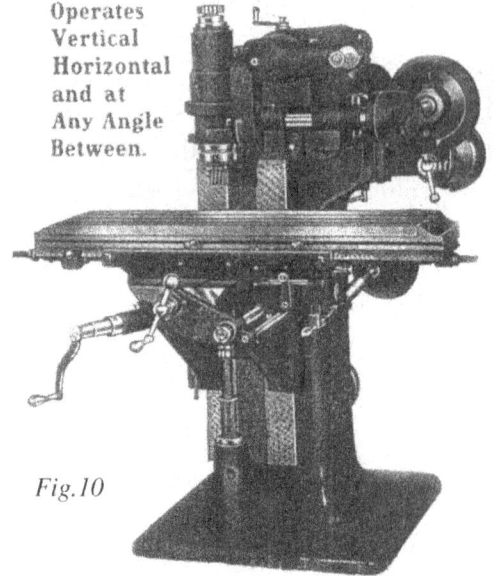

Operates Vertical Horizontal and at Any Angle Between.

Fig.10

No. A and No. AA plain milling machines (Fig.9) were introduced in 1908, followed by an improved No. 2 machine (Fig.10) and the new, constant speed No. 3 (Fig.11) in 1910. A No. 1/2 machine (Fig.12) was introduced in 1911.

In 1912, the firm reorganized as the VAN NORMAN MACHINE TOOL CO. which continued production of the DUPLEX machines.

Fig.11

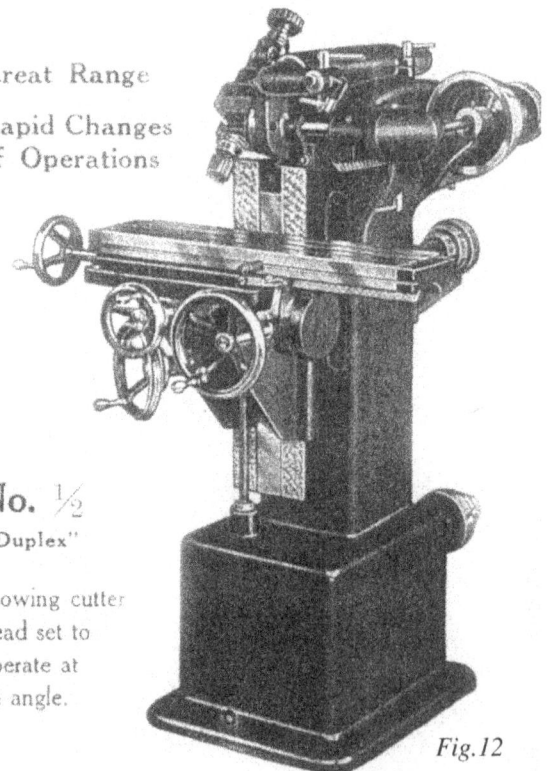

Great Range

Rapid Changes of Operations

No. 1/2 "Duplex"

showing cutter head set to operate at an angle.

Fig.12

WARNER & SWAZEY, Chicago, IL, later Cleveland, OH

later **WARNER & SWAZEY CO.,** Cleveland, OH

A partnership of Worcester R. Warner (1846-1929) and Ambrose Swazey (1846-1937) formed in 1880. Both partners had been apprentices at the Exeter Machine Works and both joined the PRATT & WHITNEY CO. in 1870 where they worked as contracting foremen.

The firm moved to Cleveland, OH, in 1882, probably because of a 140 machine order placed by a Lorain, OH, brass company. In 1900, the partners incorporated, with a capital of $500,000, as the WARNER & SWAZEY CO.

Early products centered around brassworking machines including vertical milling machines (Fig.1) "adapted to work requiring small and delicate cutters like letter-cutting or to give the rose finish to brass work" and two-spindle valve milling machines (Fig.2) weighing 1,050 pounds. Both machines were available by 1884. A four spindle valve milling machine (Fig.3), for milling four surfaces at a time, was introduced by 1900.

The only conventional milling machine built by Warner & Swazey was a Lincoln type manufacturing mill (Fig.4) introduced in 1901.

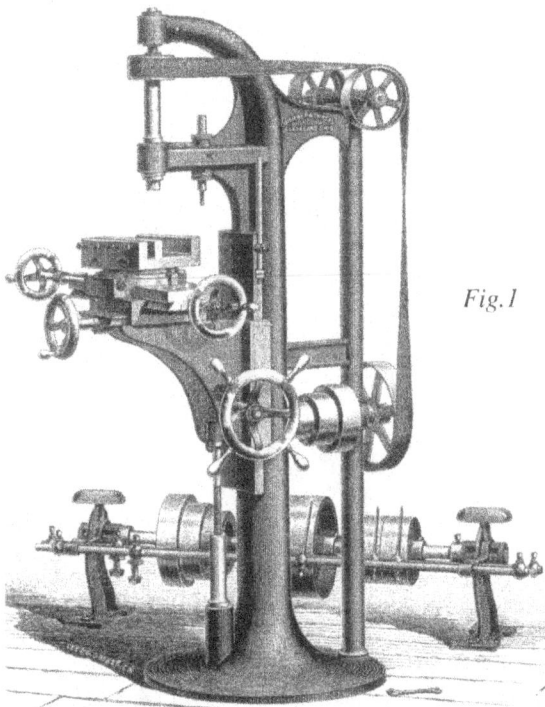

Fig.1

VERTICAL MILLING MACHINE.

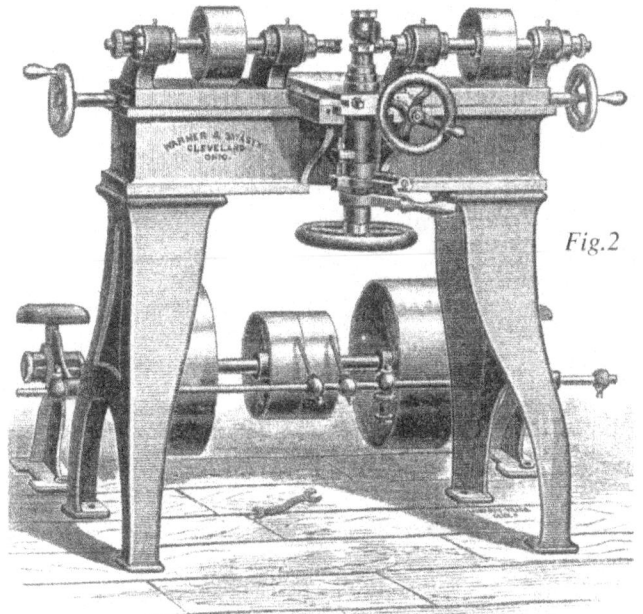

Fig.2

TWO-SPINDLE VALVE MILLING MACHINE.

Fig.3

Fig.4

WELLS BROTHERS & CO., Greenfield, MA

Formed in 1876 as a partnership of Frank O. Wells (1855-1935) and his brother Frederick E. Wells (1844-1936). In 1912, the firm merged with the Wiley & Russell Mfg. Co. to form the Greenfield Tap & Die Co.

Introduced in 1882, the only milling machine made by Wells Brothers was a power feed bench miller, *shown at right*, with 12" table travel, 3" cross travel, and 6" vertical travel.

WHEELER & CO., J.S., Worcester, MA

A partnership of Joshua S. Wheeler, Asa N. Wheeler and Orange Wheeler, formed in 1867 to make lathes, planers, upright drills, milling machines, and other machine tools. The firm operated until 1905.

WHITIN & SON, P., Whitinsville, MA

Operator of a large machine shop making textile machinery, the firm also made a variety of small machine tools for the New England fire-arms industry during the Civil War. Production is believed to have included lathes and milling machines.

NEW MILLING MACHINE.

WHITING & COMSTOCK, Hartford, CT

Maker, in 1920 of semi-universal dividing heads as *shown at left*.

WHITNEY ARMORY, New Haven, CT

later **WHITNEY ARMS CO.,** New Haven, CT

Founded in 1841 by Eli Whitney, Jr., the firm made a variety of firearms until selling out to the Winchester Repeating Arms Co. in 1888.

One of the machines taken over by Winchester in 1888 was a milling machine, shown below, still in use in 1911. The machine, probably built at the Whitney Armory, is unmarked but clearly based on the William H. Robertson patent of October 5, 1852. The Robertson patent was the first American milling machine patent.

(illustration on next page)

Whitney Armory

OLD WHITNEY MILLER AT WINCHESTER SHOPS

WHITNEY, ELI, New Haven, CT

Whitney (1765-1825) was one of the earliest contractors for U.S. military arms, beginning in 1789. A milling machine, believed to have built by Whitney in 1818 for use in arms making, was long touted as the first milling machine. The machine, *shown at right,* stands 18" high and is missing the table.

Later research has shown that the machine is probably somewhat later than 1818 and has raised doubt that it was actually made by Whitney. Nevertheless, it is a very early machine and probably the oldest still in existence.

WHITNEY MFG. CO., Hartford, CT

Formed in 1896 by Clarence E. Whitney (1869-1933) to take over the plant and business of the WOODRUFF MFG. CO. Clarence E. Whitney, the son of Amos Whitney of Pratt & Whitney Co., served as president until his death in 1933.

The firm continued production of the WOODRUFF keying system which included hand milling machines in two styles (Figs.1-2). Hand milling machines with overarms were introduced in 1901. Production consisted of a No. 1 machine (Fig.3) with 14" table travel, 6" cross travel and 15" vertical travel with another $4\frac{1}{2}$" of adjustment in the spindle head; and a No. 3 machine (Fig.4) with 14" table travel, $6\frac{1}{2}$" cross travel and 17" vertical travel with another 5" of adjustment in the spindle head. These machines, with some further improvements (Fig.5), were offered at least as late as 1920. *(illustrations on next page)*

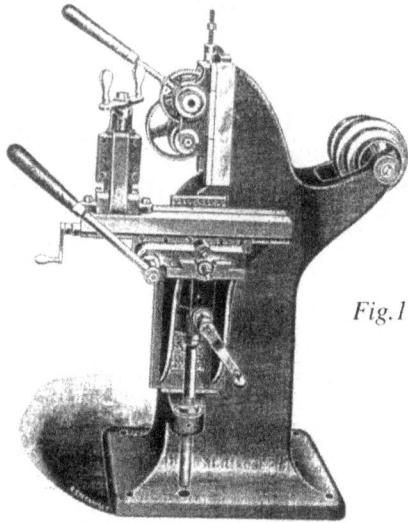

Fig.1

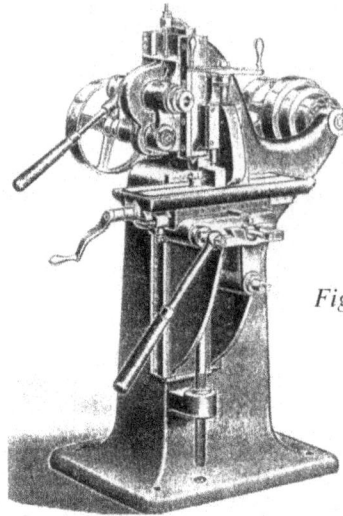

Fig.2

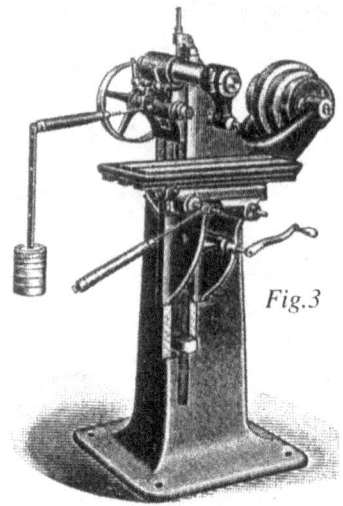

Fig.3

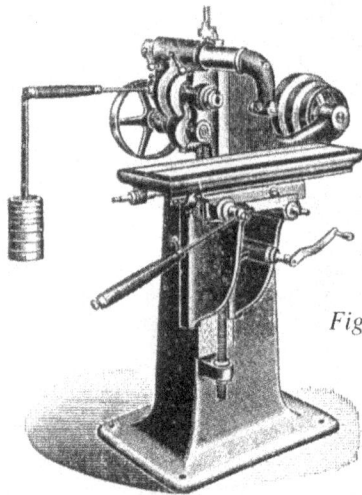

Fig.4

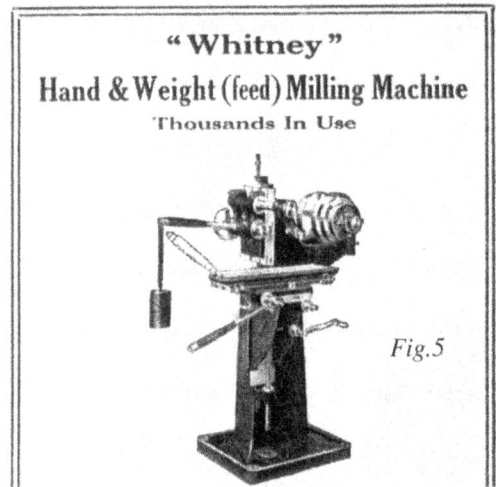

"Whitney"
Hand & Weight (feed) Milling Machine
Thousands In Use

Fig.5

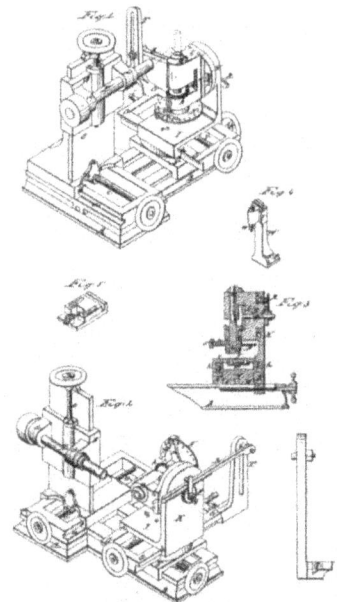

WILDER, MOSES G., West Meriden, CT

Primarily a builder of punch presses and other metal forming machinery, the firm also offered small index milling machines, patented October 21, 1862, by Wilder.

WILLARD PRESS & TOOL CO., Cincinnati, OH

Formed in 1918 as an offshoot of the Willard Machine Tool Co. Products included a 10" universal dividing head as *shown at right*. The firm appears to have failed in the early 1920s.

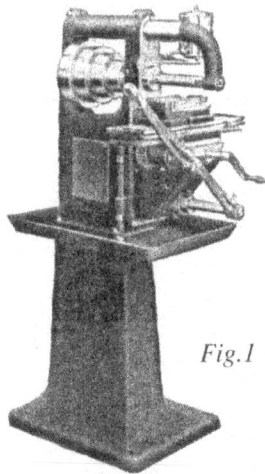

Willard 10" Universal Dividing Head

Built in
Very Liberal
Proportions

Its exceedingly liberal proportions enable it to handle the heaviest cuts encountered on a plain miller. Graduated swivel head can be set at any angle from 10 degrees below the horizontal to 10 degrees below the vertical. A large index plate, drilled on both sides, allows indexing up to 100; a chart shows divisions obtainable up to 380. Made either right or left hand. *Write for details.*

THE WILLARD PRESS & TOOL CO.,   CINCINNATI, O.

WINDSOR MFG. CO., Windsor, VT

Formed in 1865 by Ebenezer B. Lamson (1814-1892) and his son Eastborne E. Lamson as a reorganization of E.G. LAMDON & CO. Henry D. Stone continued as superintendent.

The firm built quarrying machinery and a variety of machine tools based on Frederick W. Howe and Henry D. Stone designs originally developed for ROBBINS & LAWRENCE and also made by E.G. Lamson & Co.

Machine tool production in 1865 included index milling machines and Lincoln/Howe type milling machines.

WISCONSIN MILLER MFG. CO., Milwaukee, WI

One of the many companies formed during World War I to furnish machine tools for the rapidly expanding American industrial base. As with most, it seems to have failed soon after 1918.

Products were limited to hand milling machines in cone drive (Fig.1) and electric drive (Fig.2) models.

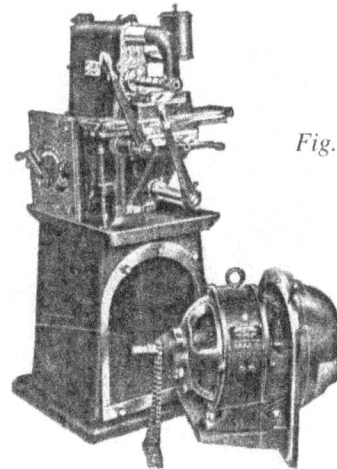

Fig.1

Fig.2

WOOD, LIGHT & CO., Worcester, MA
later WOOD & LIGHT MACHINE CO., Worcester, MA

A partnership of Aurin Wood (1812-1896) and Joseph F. Light, formed in 1852 as a reorganization of Woodburn, Light & Co. The firm reorganized as a stock company, Wood & Light Machine Co. in 1874. Light was replaced by S.S. Jennison in 1880 and the firm became Wood, Jennison & Co. *(continued next page)*

Products included index milling machines made in the early 1850's (Fig.1) and shown still in use, in 1912. The machines were advertised ca.1860 (Fig.2) for $550. Ads in 1867 offered "milling and index milling machines." By 1870, its ads no longer mentioned milling machines.

Fig.1

Fig.2

WOODRUFF MFG. CO., Hartford, CT

Formed in 1890 by William N. Woodruff (1846-1899) to make his keying system, still known as the Woodruff key.

The system consisted of half round keys (Fig.), special cutters (Fig.), and milling machines designed to cut the key seats (Fig.). The milling machine was patented January 14, 1890, by Moses C. Johnson. An improved version (Fig.) was introduced in 1893.

In 1896 Woodruff sold out to Clarence E. Whitney, who reorganized as the WHITNEY MFG. CO.

(illustrations continued on next page)

Fig.1

—Different Sizes of Keys.

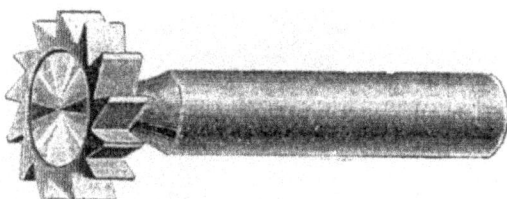

—Milling Cutter.

Fig.2

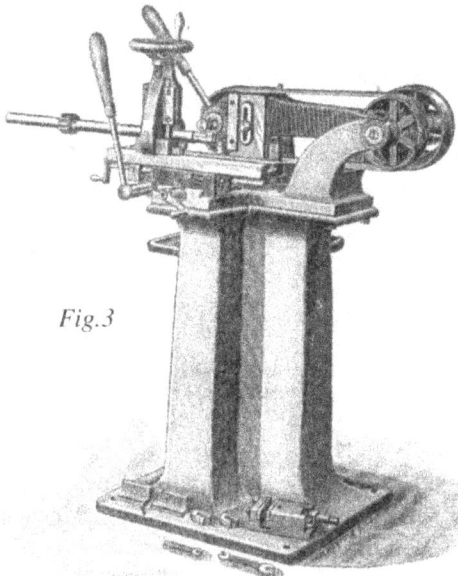

Fig.3   Woodruff

MILLING MACHINE.

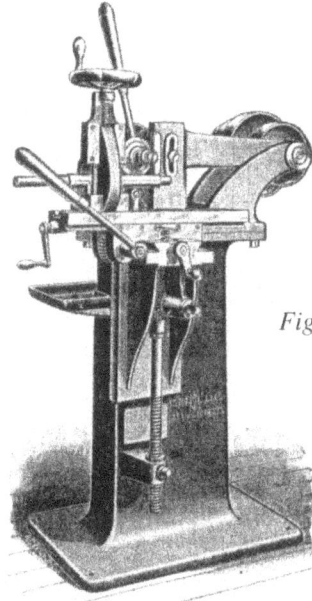

Fig.4

Hand and Key-Seat Milling Machine.

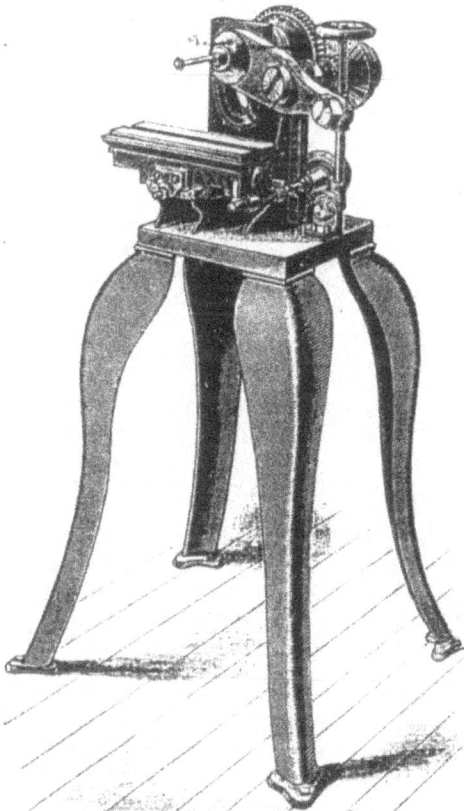

WORMWOOD, S.P., Watertown, MA

Builder, about 1885 of small power planers and milling machines. Note the unusual vertical cutter adjustment *shown at left.*

www.ingramcontent.com/pod-product-compliance
Lightning Source LLC
Chambersburg PA
CBHW081813200326
41597CB00023B/4239